HOME PLUMBING

REPAIRS AND MAINTENANCE

HOME PLUMBING

REPAIRS AND MAINTENANCE

Julian Worthington
and David Knight

W. Foulsham & Co. Ltd.

London · New York · Toronto · Cape Town · Sydney

W. Foulsham & Company Limited
Yeovil Road, Slough, Berkshire, SL1 4JH

ISBN 0-572-01262-4

Copyright © 1984 W. Foulsham & Co. Ltd.

All rights reserved.
The Copyright Act (1956) prohibits (subject to certain very limited exceptions) the making of copies of any copyright work or of a substantial part of such a work, including the making of copies by photocopying or similar process. Written permission to make a copy or copies must therefore normally be obtained from the publisher in advance. It is advisable also to consult the publisher if in any doubt as to the legality of any copying which is to be undertaken.

Photoset in Great Britain by C. R. Barber & Partners (Highlands) Ltd., Fort William, Scotland, and printed in Hong Kong.

We should like to thank the following manufacturers for their help in illustrating this book with the generous loan of tools, equipment, fittings and some product photographs.
    Deltaflow Ltd., Hunter Building Products Ltd., Ideal-Standard Ltd., Kay & Co. Ltd., Key Terrain Ltd., Marley Extrusions Ltd., IMI Opella Ltd., The Sylglas Company, Torbeck Control Valves Ltd., Uni-Tubes Ltd., Walker Crosweller & Co. Ltd. and Yorkshire Imperial Fittings.
    We should also like to thank J. H. Sutton & Son Ltd. of Marden, Kent for the loan of plumbing tools and equipment.

Contents

Introduction — 7
Local authority regulations — 8
Electrical earth return — 8
Old pipe systems — 8
Company stopcock — 9
Corrosion — 9
Dezincification — 9
Tap labelling — 10
Pipe sizes — 10

1 Domestic systems — 11
Cold water distribution — 11
Hot water distribution — 14
Waste system — 17

2 Tools and accessories — 19
Routine maintenance — 19
Installation work — 20

3 Basic maintenance — 24
Repairing dripping taps — 24
Repairing leaking taps — 27
Repairing a cistern valve — 29
Replacing a WC cistern diaphragm — 33
Freeing a jammed stopcock — 35
Draining the system — 36
Repairing leaking joints — 36
Lagging pipes — 38
Lagging a cistern or cylinder — 39
Repairing burst pipes — 41
Clearing a blocked pipe — 43

4 Metal components — 45
Copper pipe — 45
Stainless steel pipe — 45
Other metal pipes — 46
Bending pipe — 46
Capillary fittings — 48
Compression fittings — 50
Special fittings — 52
Taps and valves — 55

5 Plastic components — 58
PVC tubes — 58
Polybutylene tubes — 58
Polyethylene tubes — 60
Solvent-weld joint — 60
Mechanical joint — 63
Plastic taps and fittings — 65
Plastic pipe clips — 66

6 Waste fittings — 67
Overflow pipe — 67
Sink, wash-basin and bath waste pipe — 68
Soil pipe — 70
Joining waste pipes — 74
Traps — 76

7 In the kitchen — 80
Fitting a mixer tap — 80
Replacing a sink — 84
Plumbing in a washing machine/dishwasher — 88

8 In the bathroom — 91
Replacing a tap — 91
Fitting a mixer tap/shower unit — 92
Replacing bathroom fittings — 94
Replacing the WC — 96
Replacing a bath — 100
Replacing a wash-basin — 103

9 Adding to the system — 105
Connecting to the stack pipe — 105
Fitting an outside tap — 107
Installing a shower unit — 108
Installing a bidet — 114
Installing a bedroom vanity unit — 115

Index — 118

Introduction

Water is without doubt the most essential service to any home. This is reflected in the fact that today the requirement of water per person is estimated to be in the region of 200 gallons (900 litres) per day.

If this seems excessive, you must bear in mind that the average bath uses 20 gallons (90 litres), washing takes about five gallons (23 litres) and every time you flush the WC you use two gallons (9 litres).

When you add to these basic requirements the amount of water consumed in preparing and cooking food, washing up, general house cleaning and washing clothes, it does not take long to reach this apparently surprising figure.

With this amount of water being used by a single person every day, plumbing installations have of necessity become more complex.

You only have to look at the number of fittings in the average home – kitchen sink, bath, wash-basin, WC, boiler, central heating and possibly shower and bidet – and water-consuming appliances such as washing machines and dishwashers to realise that a relatively complex system has to be arranged to cope with all these items.

As the number of installations in the home increases, so does the need for maintenance and general repair work – as well as a basic understanding of the domestic plumbing system. In these days of high labour costs, it is expensive to call in a plumber for every small job that arises.

With this in mind, this book is designed to familiarise you with the basic plumbing installations in the average home and to describe the common maintenance and repair jobs you are likely to come across and what you will need to carry them out.

The book also shows you how to install the basic fittings in the home, a task that should be well within the scope of the competent DIY person, particularly since there is now a whole range of plumbing-in kits available to keep the work as simple as possible.

However, the book does not cover such items as central heating installation or below-ground drainage, since these call for more specialised knowledge and skills, which you will not be in a position to handle with any confidence until you have mastered the techniques included in this book.

The makers of plumbing fittings are increasingly aware of the amount of DIY work now undertaken on the domestic system and are responding to this demand by producing pre-packaged kits of components for many of the jobs around the home, such as plumbing in a washing machine.

Many new products have been introduced to help the DIY enthusiast undertake much of the work that previously required a high degree of knowledge and skill. Lead and steel pipework has now been superseded by copper, which is much easier to work with, and plastic materials – with their even simpler application – are being approved by the water authorities for more and more uses. One major advantage of plastic nowadays is that some products are approved for use with hot, as well as cold, water.

Because of the continual advances being made in technology, new products and processes are coming onto the market all the time. When you come across these, it is important to check that the items involved have been approved by the water authorities and do not contravene specific local regulations.

Local authority regulations

Because of the geological differences from area to area, the make-up of the domestic water supplied to your home will vary accordingly.

Of course water supplied anywhere in the United Kingdom is sterile and safe to use, but you will find in certain parts of the country that dissolved calcium and magnesium salts make for hard water, whereas in other parts the water is soft.

In general, hard water will deposit scale if it is heated to temperatures in excess of 71°C, whereas soft water will not. This and other chemical differences between water supplies in different areas means that regulations will also vary to cope with the local situation.

Another factor affecting the regulations is the amount of water stored in individual reservoirs. For example, where there are massive underground storage facilities, there is normally a more liberal attitude to the use of water than in areas where the storage capacity is limited.

These, and other considerations, will determine what is or is not acceptable as regards home plumbing installations in any given area. This is why you must always check with your local water authority before you undertake any new installation on the system to ensure that what you intend doing is in line with current regulations.

A point to remember here is that you will not normally need permission to replace an existing fitting with a similar new one, such as a bath or WC.

One important consideration in terms of regulations is that of safety, which applies wherever you happen to be living. It is essential that the main supply of water is protected at all times from contamination by water that has been used or stored within the domestic system.

For this reason all stopcocks fitted should be of the 'non-return' type and special measures have to be taken when installing such items as mixer taps, bidets and so on. These are included in the relevant sections of the book.

Electrical earth return

In some areas the earth return for all the domestic electrical installations is via the metal water pipes buried in the ground. This method of earthing is normally visible by an earth wire connected to a clip on the main water supply pipe before the main stopcock – often sited under the kitchen sink.

With any installation where you are intending to replace metal pipes with plastic ones, you must check that there are no earth clips fitted to the affected section of pipework. Plastic is an extremely good insulator and would eliminate the safety factor of an earth return. If this is fitted to a metal pipe you want to replace, you should consult a qualified electrician or your local electricity board before attempting any alteration.

Old pipe systems

Many older houses were originally plumbed in using just lead or galvanised steel pipes. This type of system requires a high degree of skill and special tools, such as stocks and dies, when any modifications are needed. For this reason these systems are not discussed in any great detail in the book.

Special fittings are available to convert from the steel pipework, and normally you will find that lead pipes terminate with a suitable male or female thread to which you can add new pipes.

In some areas, however, there is a risk of corrosion taking place due to a reaction between certain combinations of metal. In this case special fittings must be used when joining new pipes to the old ones (see 'Corrosion' below).

Company stopcock

It is at this point on the domestic plumbing system that the responsibility of the local water authority for the system ends and the householder's begins. Normally this stopcock is located in front of the house – often in the pavement but, particularly in rural areas, it could be elsewhere near the building.

With older properties, it may be that the stopcock serves the supply to more than one house and in this case you must first check with your neighbours and prepare them for any inconvenience caused while the water supply is cut off.

The stopcocks themselves do vary in design. Some are fitted with the standard T-shaped handle, while others have a square shank spindle that can only be turned with a special key.

Since the stopcock will be located some way down in the ground to prevent the risk of possible frost damage, you will need a long length of timber with a groove cut in the end to operate the T-shaped handle. With the other type of stopcock, you will have to get in touch with your local water authority about its operation.

Corrosion

Everyone knows how water will cause steel to corrode – to produce rust. The best way of preventing this from happening is to cover the steel surface with a thin coat of zinc.

This process is known as galvanising and is used on all steel components that are in contact with water in the plumbing system. Such components include pipes, cold water storage tanks and, in some older houses, hot water tanks as well.

The use of copper pipes in domestic systems should in most cases present no problems, particularly in hard water areas. Where the water is relatively soft and possibly very slightly acidic, it can act as an electrolyte and cause a reaction between zinc and copper. As this reaction removes the zinc, the steel is left exposed to the water.

In areas where this is likely to happen, the use of the two metals in combination is discouraged and it is better to fit either a plastic storage cistern or plastic tubing.

Dezincification

This is really an extension of the corrosion problem already mentioned, but it is usually confined to the brass fittings used with copper pipe or plastic tube. Again it only applies to certain areas where there is soft water.

The fittings normally described as brass are made from a mixture of copper and zinc. In problem areas the water acts on the alloy in such a way as to remove the zinc. This is either dissolved in the water and washed away through the system or it is deposited elsewhere in the system, where it may cause blockages. The fitting itself ends up as a spongy, porous mass of copper which can leak and in extreme cases fracture.

During the 1970s, much research was carried out into the problems involved and the result was the approval of a different alloy that could resist this type of attack. Fittings made with this special alloy are classed as 'dezincification resistant' and are marked with a special symbol – CR.

Unfortunately these fittings are more expensive, but should be used in all plumbing installations – both above and below ground – in areas where this problem exists. You should check, also, that any taps, valves, mixers, etc fitted are also resistant to attack.

Tap labelling

Whenever you remove taps from their supply pipes, you should label each pipe 'hot' or 'cold' to remind you which supply is which when you come to refit the taps later. You will not always remember and it is very difficult to tell the two apart when the hot pipe has cooled down.

Pipe sizes

All modern fittings sold today are metric, but you will come across the old sizes and terminology being used and this can be confusing.

The size of steel pipe is given as the 'nominal bore'. Since this used to be the system most commonly used, many fittings are still referred to with reference to the thread sizes used with this pipe.

Thus a $\frac{1}{2}$in tap has a threaded tail on it with a $\frac{1}{2}$in British Standard Pipe (BSP) thread on it. This thread has an external diameter of approximately $\frac{7}{8}$in. A $\frac{3}{4}$in tap has a $\frac{3}{4}$in BSP tail, the thread of which has an external diameter of about $1\frac{1}{16}$in.

Fittings that connect to steel pipe or taps have the same thread sizes. These are often referred to as 'iron', 'barrel' or 'pipe' threads.

The following table gives approximate conversions used to relate with copper pipe and plastic tube sizes. The sizes given represent the nominal size of the pipe and do not necessarily refer to the specific bore or external diameter of the pipe.

Type of pipe	*Metric size*	*Imperial size*
Copper* or stainless steel water pipe	15mm	$\frac{1}{2}$in
	22mm	$\frac{3}{4}$in
	28mm	1in
	35mm	$1\frac{1}{4}$in
	42mm	$1\frac{1}{2}$in
PVC cold water tube	15mm	$\frac{1}{2}$in
	22mm	$\frac{3}{4}$in
	28mm	1in
PVC overflow pipe	22mm	$\frac{3}{4}$in
Polybutylene & polyethylene tube	15mm	$\frac{1}{2}$in
	22mm	$\frac{3}{4}$in
	28mm	1in
PVC and polypropylene waste pipe	32mm	$1\frac{1}{4}$in
	40mm	$1\frac{1}{2}$in
	50mm	2in
PVC stack pipe	82.4mm	3in
	110mm	4in
	160mm	6in

*The old Imperial size copper pipe is not always interchangeable with the modern metric pipe (see page 45).

1 Domestic systems

Cold water is fed from the nearest reservoir to your immediate area through large diameter pipes, which are made of cast iron, pitch fibre or other suitable material. A branch from this main then runs to each property.

The supply into each property is controlled by a special water authority stopcock, which is usually situated in front of the building beneath a small hinged metal plate. The stopcock itself will be at least 1m (or 3ft) below ground and may have a special shank that can only be turned by a particular key, which the water authority will hold.

With some older properties the authority stopcock isolates more than one property. If for any reason you have to turn it off, first check that no-one else will be unduly inconvenienced.

From this stopcock a 15mm (or $\frac{1}{2}$in) nominal bore pipe, which may be made of galvanised steel, lead, copper or plastic and is usually protected from possible subsidence by sections of drain pipe, runs on a slight upward incline into the property.

The incline is important to ensure no air bubbles remain in the section of pipe between the main and the inside. At no point before it enters the building should this pipe be less than 750mm (30in) below the surface, since it could otherwise be affected by frost.

Cold water distribution

Once under the walls, the supply pipe rises up through the floor and here the main stopcock is connected. This valve allows water to pass through in only one direction; this is to prevent any contaminated water running back into the mains supply in the event of a pressure failure. It also enables you to cut off the cold water supply into the home, which you may need to do in an emergency or when working on certain parts of the system.

Immediately above the main stopcock you will find a draincock fitted. You will only need to use this if you have to drain the rising main for the purposes of maintenance or other major work on the system.

From this point, systems do vary according to the local water authority regulations. Many authorities will only allow the cold water tap at the kitchen sink and one outside tap to be fed direct from the main pipe and will insist that all other taps and relevant fittings are fed from the cold water storage cistern.

The reason for this is that the demand for water fluctuates through the day, and is usually heaviest in the early part of the morning. If every tap and appliance, including the WC cistern, was connected directly to the mains, the demand could well exceed the supply available at any one time. In some circumstances this could result in contaminated water flowing back down the mains pipe.

By feeding most of the water outlets in the home through a storage cistern, a sufficient supply of water at any time is guaranteed and the cistern can refill at a lower usage rate until the demand falls off again.

Some authorities, however, do allow you to feed the cold water tap on the wash-basin and the WC cistern direct from the mains – and occasionally the bath. You will have to check on this with your local water authority.

The cold water tap at the kitchen sink will normally be the first branch after the draincock above the main stopcock, from which a garden tap may also be fed, although regulations vary.

The rising main then runs up inside the house to the cold water storage cistern, which is usually situated in the loft space. This cistern is located as high as possible in the house to ensure a good head (or pressure) at all the outlets fed from the cistern.

The size of the cistern varies according to regulations, but the capacity for the average home is 50 gallons (227 litres). Because this will weigh more than 500lb (227kg) when full, it is usually situated over a supporting wall. If possible, it should be sited against a warm flue or chimney – and certainly away from the eaves to minimise the risk from frost damage.

While on this subject, the cistern itself should be insulated for protection from the cold weather but you should never lay loft floor insulation immediately under it. The reason for this is to allow whatever warmth comes up from the room below to reach the cistern.

Cold water storage cisterns used to be made from galvanised mild steel. Nowadays they are all made of glass reinforced or other plastic.

These types have distinct advantages over their predecessors. Not only are they lighter and easier to install in the loft space, but they are much simpler to work on when connecting fittings to them. But the most important advantages are that they are completely resistant to corrosion and have better thermal insulation properties, which means the water inside is less likely to freeze up in very severe conditions.

The storage cistern acts as a reservoir from which the outlets throughout the home draw their supply. The water enters from the rising main through a float-controlled valve, which regulates the flow of water into the cistern and maintains it at a predetermined level.

In the event of the valve failing to cut off the water completely when this level is reached, there is a warning pipe running on a slight downward incline from just above the recommended water level to the outside of the house. Should the water rise above the specified level, it will run away along this pipe and prevent the possibility of the cistern overflowing. The warning pipe will also provide you with a visible sign of a fault in the cistern valve.

The outgoing supply, to feed outlets through the home, is usually through two pipes fitted just above the base of the cistern. They are fitted here to prevent the possibility of any sludge or scale deposited in the bottom of the cistern getting into the water supply system.

One of these pipes feeds all the cold water outlets (apart from the kitchen sink tap and possibly outside taps) in the home, such as the bath, wash-basin, bidet, WC cistern, etc. The other runs to the bottom of the hot water storage cylinder to supply water to that system.

Normally you will find gate valves fitted to each pipe near the cold water storage cistern. This is to enable you to isolate the water supply when doing maintenance and repair jobs to the system, such as changing a tap washer, without having to empty the storage cistern first.

If you have a hot water central heating system installed in your home, you will find there is another, smaller storage cistern next to the main one, which is also fed from the rising main. This is known as the feed and expansion tank, whose job it is to supply water to the boiler and radiators. This cistern, which usually has a capacity of 10 gallons (45 litres), has its own float-controlled valve and warning pipe, which operates in the same way as on the main cold water storage cistern.

It is important to note here that this warning pipe should also feed to the outside of the house and not over the larger cistern, as is sometimes

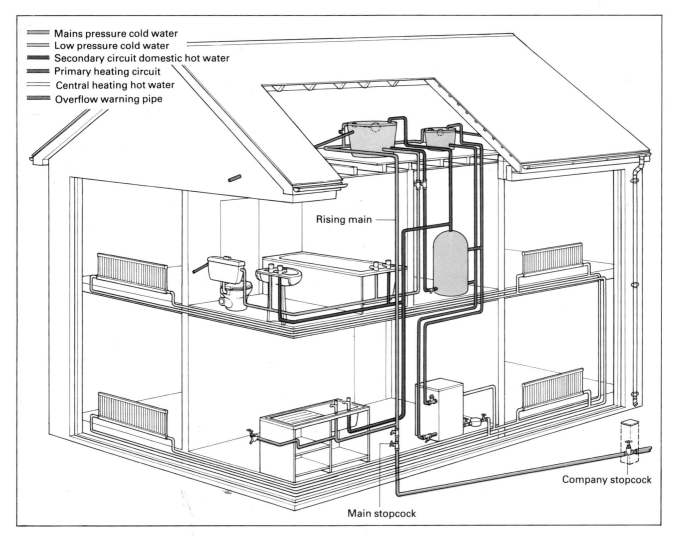

This example of a typical domestic plumbing system will help you understand how the hot and cold water supplies are fed round the home. From the diagram you should be able to locate the various control valves and draincocks on the hot and cold systems. The hot water system shown in this illustration is of the indirect type and includes a central heating system via a series of radiators.

the case. Normally additives are contained in the central heating system water to help prevent corrosion or the formation of scale. If the expansion tank overflows, this contaminated water must not be allowed to drain off into the main domestic cistern.

Hot water distribution

The method of distribution of hot water through the home will depend on whether or not you have a central heating system installed, as well as just a domestic hot water system, and whether the system is direct or indirect.

Direct system With this type of system, which may exist in older properties but is now being replaced by the indirect system, the hot water is stored in a copper cylinder of between 25 and 35 gallon (114 and 160 litre) capacity or, in even older systems, in a sealed galvanised steel tank. The water can either be heated by a boiler or an immersion heater.

As already mentioned, a feed pipe runs from the bottom of the cold water storage cistern to the bottom of the hot water cylinder. From the top of this cylinder a vent pipe runs up and terminates in an open 'U' bend over the storage cistern. And it is from this rising vent pipe that the supply is taken to all hot taps in the home.

If a boiler is used to heat the water, a feed pipe is taken from the bottom of the hot water cylinder to the boiler and a return pipe from the boiler then runs back to near the top of the hot water cylinder. In most direct systems large – usually 28mm (or 1in) nominal bore – pipes are used and the flow is created by convection, with no pump fitted to the system.

A safety valve is incorporated into the system near to the outflow at the top of the boiler and a

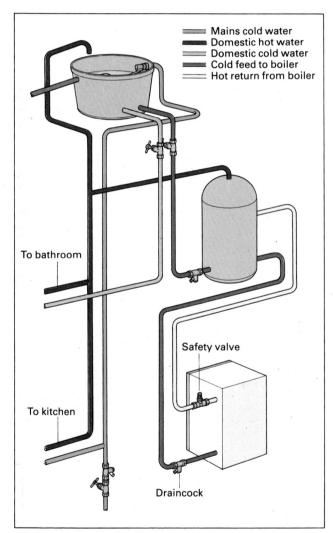

This diagram shows how a direct hot water system is plumbed in, incorporating a standard cylinder and boiler. The hot and cold water feed pipes are also indicated.

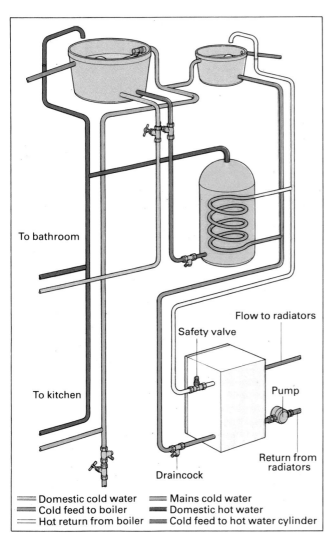

With the indirect system, the primary circuit water is sealed in its own system and feeds the central heating, while the secondary circuit supplies the hot taps around the house.

- ═══ Domestic cold water
- ═══ Cold feed to boiler
- ─── Hot return from boiler
- ═══ Mains cold water
- ═══ Domestic hot water
- ═══ Cold feed to hot water cylinder

draincock is fitted at the lowest point of the inflow to the boiler, should you need to empty the system at any time.

If your system uses an immersion heater, this will be fitted inside the hot water storage cylinder or tank. There are basically two types of immersion heater. The single element type heats the whole cylinder or tank, while the twin element (bath or sink) type can be used to heat just part of the water in the cylinder or tank. Here there is one short and one long element and you can switch on just the short element to heat the top 250mm (10in) or so of water.

There used to be a compact version of this system, available as a packaged unit, which was oval in shape to take up less space. The unit comprised a small storage cistern fed from the main storage cistern and a 25 gallon (114 litre) cylinder immediately beneath it. Larger units are currently available, where local regulations permit. These are not suitable in hard water areas where there is a risk of scale formation.

The major disadvantage with the direct type of system is that you cannot use corrosion inhibitors because the hot water in the system is constantly being drawn off through the hot water taps. The indirect type incorporates a primary circuit, which is sealed and from which hot water is not being drawn off. So here inhibitors will remain in the system's water.

In hard water areas the dissolved calcium bicarbonate in the water is changed into insoluble calcium carbonate at temperatures above 71°C. This is then deposited as scale on the inside of the boiler, hot water cylinder and pipes. The rate of deposit is high and it is quite possible for a 28mm (or 1in) diameter pipe to become totally blocked by scale. Boilers become very inefficient due to the scale's insulation effect and the walls overheat leading to failure.

Indirect system With this type of system, the problems associated with corrosion and scale formation are minimised as the water passing through the boiler is never drawn off and only those losses through evaporation are made up with fresh water. When this water is initially heated the dissolved oxygen, which is a major contributory factor in corrosion, is driven out and so the scale deposits that do form are very small and cannot be added to.

You can get corrosion inhibitors to add to the water in the primary circuit, thus providing virtually complete protection to the boiler and the hot water cylinder.

This primary circuit can be extended to include radiators as well, which means similar protection for the central heating system. The primary circuit starts at the small storage cistern in the loft. Here the ball valve is adjusted to shut off the mains supply when the water is 25–50mm (1–2in) deep when cold. When the system heats up, the water will expand and the level in the expansion tank will rise, but not as far as the warning pipe.

The feed from this cistern is taken from the bottom to the flow entry to the boiler. From the boiler, a vent pipe runs up and terminates in an open-ended 'U' bend over the expansion tank. The feed to the hot water storage cylinder and the radiators, where central heating is installed, is taken from this vent pipe.

When water from the boiler enters the hot water cylinder, it is fed to a sealed calorifier coil inside (see below). The return flow from the calorifier and radiators is fed back to the boiler.

Normally a pump is fitted to the system to circulate the hot water through the radiators, which are supplied via 15mm (or ½in) small bore pipes. The feed to and return from the calorifier are always through large bore pipes – normally 28mm (or 1in) – and work on the principle of convection, as with the direct system.

The hot water cylinder used with an indirect system is of a special type that incorporates a calorifier, a copper coil inside the cylinder that is surrounded by the domestic water. The feed from the boiler enters at the top of the calorifier and, due to the cooling effect of the surrounding domestic water, flows back from the bottom of the calorifier to the boiler.

With most boilers, the central heating circuit with its pump is fitted to one side of the boiler and the domestic hot water circuit is taken from the other side.

The secondary circuit supplies the hot taps around the home. The water is fed from the cold water storage cistern to the bottom of the hot water cylinder via a pipe on which a gate valve is normally fitted near the bottom of the

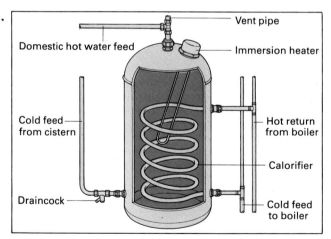

The indirect type hot water cylinder as shown here incorporates a calorifier, through which the primary circuit water flows. The hot water for taps is stored around the calorifier.

storage cistern. The domestic water fills the cylinder and surrounds the calorifier.

From the top of the cylinder a vent pipe runs up and terminates in a 'U' bend above the cold water storage cistern. The feed to the hot water taps in the house is taken from this vent pipe.

Waste system

Having looked at the way hot and cold water are distributed around the home, we need to see where the water goes after it has been used. Depending on where your home is situated, this will be to the main drainage system or to a cesspool and it gets there in one of two ways.

With houses built before the mid-1960s, waste water was removed through the two-stack system. Since then a single-stack system has been introduced which is incorporated within the building to minimise the risk of frost damage, reduce costs and simplify the pipework.

Two-stack system As the name implies, this type has two cast-iron or plastic waste pipes. The larger diameter pipe carries the waste from the WCs and is called the soil pipe. This is connected directly to the drains and extends above the tiles on the roof to act as a vent to release sewer gasses.

Any ground-floor WC will normally be connected directly to the main drains via a short branch pipe, while any WC above this level will be situated near the soil pipe and be linked to it via a short connecting pipe.

The second stack pipe, which is smaller in diameter, takes the waste from the other fittings, such as the bath and wash-basins. The waste from the sink may be connected to it or run directly to a gully outside the house. Like the soil pipe, the waste pipe also runs down the side of the building, usually starting above the tiles and finishing in a yard gully. This gully is connected to the main drains via a special trap.

Older baths and wash-basins normally had copper or brass traps fitted, which had shallow water seals (see pages 77–79). In the mid-1960s, a change in building regulations required all drainage stacks to be contained within the fabric of the house, using the single-stack system. There are, however, properties where a single-stack system exists outside the building.

Single-stack system This incorporates a single large soil and vent pipe of at least 110mm (or 4in) diameter. Originally this was made from cast iron, but muPVC pipe has replaced it.

The soil and vent pipe is connected to the main drainage system via an easy bend to minimise the risk of blockages and extends to above the tiles. In common with the two-stack system, the top of this pipe has a wire cage over it to prevent birds, leaves, etc getting inside.

The connections to this pipe must be as short as possible. This is to prevent the possibility of the water-seal in waste traps being siphoned out should the waste pipe run full. For pipes of 32mm (or 1¼in) diameter the maximum distance allowed from the trap to the stack pipe is 1·68m (or 5½ft). If the length of pipe run is greater than this, the trap must be vented.

With the single-stack system, deep seal traps must be used on fittings and these should be of the 'P' type. WC connections should be as close as possible to the stack pipe and 'swept' fittings (see page 72) must be used to minimise the possibility of blockages.

Because of these restrictions on the system, modern houses are now usually designed with the kitchen and bathroom either next to one another or with the bathroom above the kitchen.

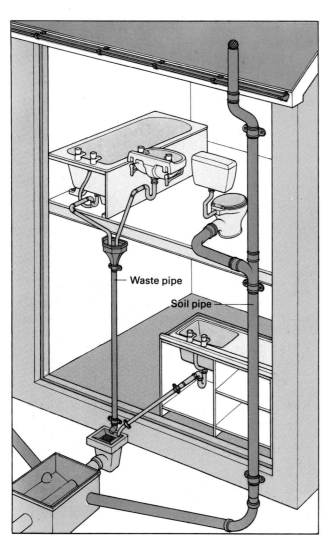
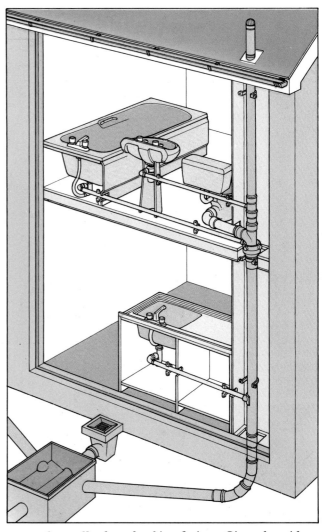

Depending on when it was built, your home will have either a single-stack or two-stack waste system. With the two-stack system, the larger stack pipe carries the waste from the WCs, while the narrower one takes the waste from all other plumbing fittings. Since the mid-1960s, the single-stack system, usually contained inside the house, has been used. This involves simpler pipework, costs less and minimises possible frost damage.

2 Tools and accessories

As with any other aspect of DIY work, it is very important to work with the right tools. Apart from making the job more difficult, the use of inappropriate tools can also cause damage to the various plumbing joints and fittings. It is equally important that the tools you use are in good condition and not damaged in any way.

The first section lists the type of tools required when you are tackling general maintenance work on the system – and here you may find you already possess some of those needed. The second section covers the type of tools you may want when you come to look at installation work on the system.

Routine maintenance

These tools will be useful for general repair and maintenance work on the plumbing system and fittings. You can buy what you need as and when you have to tackle specific jobs, although it is worth having the basic tools to hand in case of an emergency.

Screwdrivers You will need a reasonable size screwdriver – preferably with at least an 8mm ($\frac{3}{8}$in) blade – to undo the screws on the top of WC cisterns, where this type of fitting is used. The handle or head of a tap is often held in place with a small screw and here you will want a smaller screwdriver – ideally of the type used by electricians.

Check on the type of screws used on plumbing fittings. They could have Pozidriv heads, in which case you must have the proper Pozidriv screwdrivers. Do not get the now obsolete Phillips head screwdrivers, since these will damage the screws. As a general guide, Pozidriv screwdrivers are never chromium-plated.

Pliers You will find a sturdy pair of 150mm (6in) pliers invaluable for a whole range of jobs, such as closing or removing split pins in cistern valves. You will get a better grip with the electrician's type; these have plastic-coated handles.

Spanners There are hexagonal or octagonal nuts on all the traditional plumbing joints and many of the fittings and it is most important that you use the right size spanner for each. You may need to double up on some sizes because, as with compression fittings, one spanner is used to stop the fitting rotating while the second is used to adjust the nut – either to tighten or loosen it.

Unfortunately buying the spanners you need can be quite expensive, particularly since you will want a range of sizes. You certainly do not want to buy two of each. Ideally, the correct size spanner is the best tool to use on nuts. But to save money it is worth investing in a good quality 250mm (10in) adjustable spanner.

When you buy an adjustable spanner, make sure the jaws are parallel – and do not use one if its teeth are worn. The easiest type to use is the one that looks like a standard open-ended spanner.

A word of warning here. Never be tempted to use a Stillson or self-grip wrench on hexagonal nuts. If you do, the chances are you will ruin the nuts and leave deep marks on the soft brass of the fitting. And if the nuts are very tight, these wrenches are likely to wear down the corners of the hexagon. When this happens, you will have great difficulty moving the nut in future.

Self-grip wrench This type of wrench has many uses, since it can grip circular sections and other irregularly shaped fittings. But it must be handled with care. Its hardened steel jaws grip very tightly and hence they can easily collapse copper tube and distort brass fittings. The answer is to use the self-grip wrench only where other tools are not strong enough to do the job.

Small hammer This is a general purpose tool you may find handy from time to time. Sometimes the head or handle of a tap is difficult to move and here the hammer can be useful to loosen it. In this case, you must use small pieces of wood as packing round the tap to prevent the hammer damaging the surface. Bear in mind that many plumbing fittings are made of brittle cast metal or plastic and you should not use a hammer on these.

Packing and jointing materials On occasions you will need to repack certain types of tap (see pages 27–28). One suitable material is tallow cotton, although there are other proprietary packings on the market.

If, for any reason, you have to work on a joint, you may need to reseal it – and here you should use a jointing compound, if there is evidence that it was used when the joint was originally assembled. The liquid compounds, such as boss white, can be used to seal flat surfaces. With threads, PTFE jointing tape has largely superseded the liquid type sealers.

Installation work

You will need extra tools if you want to add to the system, replace fittings or tackle other major plumbing jobs around the home. It is best to buy these as you need them or hire them.

Stillson wrench This is best used on steel piping where you are unscrewing or tightening up, since it is ideal for gripping pipe and circular fittings. It should only be used on harder metals such as steel since it will leave teeth marks. The most practical size is 350mm (14in).

Compression joint spanners You can save yourself a lot of time and money if you are handling a large number of compression fittings by getting a special spanner to fit a range of octagonal nuts. Bear in mind CR fittings (see pages 9–10) have hexagonal nuts and these can be loosened or tightened with a normal open-ended spanner.

Hacksaw The junior hacksaw will handle all cutting work on copper pipe and plastic tube. Get one with replaceable blades so that you can change these as they wear out or break. Always ensure that you cut the pipe or tube square, otherwise you will have problems when making the connections.

File When you cut pipe or tube, you will get a burred edge on the sawn section. To smooth this down you will need a second-cut flat or half-round file – the ideal size is 200mm (8in). Make sure you buy a wooden handle to make working the file easier and safer. Files come without handles.

Pipe cutter This is a much easier way of cutting copper pipe accurately than with a hacksaw. The hardened steel cutting wheel ensures a clean, square cut. Most types of pipe cutter incorporate a tapered reamer, which you use to clean out the cut end of the pipe. So the cutter does the job of two tools.

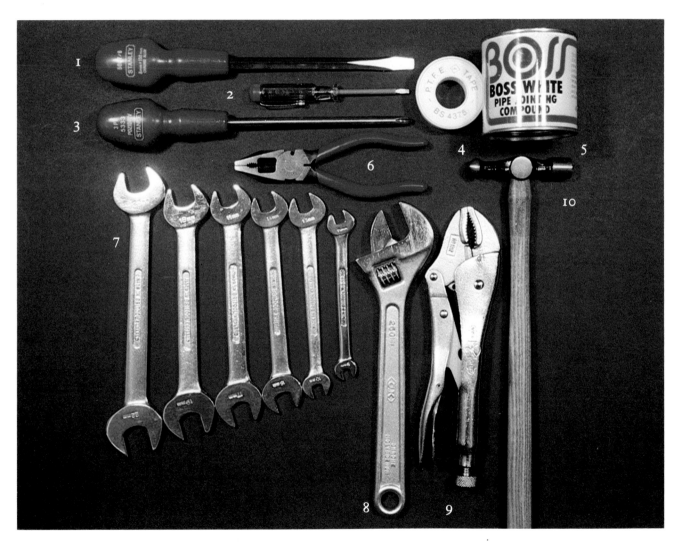

This range of tools will be useful for routine maintenance work on the domestic plumbing system: 1 Large screwdriver, 2 Small screwdriver, 3 Pozidriv screwdriver, 4 PTFE tape, 5 Boss white, 6 Electrician's pliers, 7 Set of open-ended spanners, 8 Adjustable spanner, 9 Self-grip wrench, 10 Small hammer. Some of these items you are likely to have already in your ordinary toolbox.

Bending spring This is essential if you are bending copper pipe, since it prevents the pipe collapsing or creasing at the bend. Springs come in different sizes to fit the various diameters of pipe. When using a bending spring, you will find a length of strong cord handy. If you tie the cord securely to the loop at the end of the spring, you will be able to pull out the spring when making a bend in a long length of pipe. Make sure you keep the spring well greased. If it gets rusty, it will be difficult to remove.

Pipe bending machine You should find it relatively easy to bend 15mm (or $\frac{1}{2}$in) diameter pipe over your knee – and with a little more strength you can also handle 22mm (or $\frac{3}{4}$in) diameter pipe in the same way. But with larger diameter pipe or if you have a number of bends to make, you will find the pipe bending machine much easier. The chances are that in this situation you will be tackling a one-off job and you can therefore hire the bending machine.

Apart from saving effort, the main advantages of this piece of equipment are that all the bends will be uniform and shaped to the correct angle and you will not run the risk of damaging the pipe.

Blowtorch If you are working on capillary fittings, you will need a blowtorch. The butane gas type is much more convenient and safer to work with than the traditional paraffin one. You can also use the blowtorch to anneal copper pipe, making it easier to bend.

Soldering materials You will need these if you are working with capillary fittings. You may have to use cored soft solder to 'top up' a capillary joint and you will want to apply flux to each joint before you assemble it. Before you apply the flux, you must clean up the joint – and you will need steel wool or fine glasspaper.

Basin and bath wrench You will find this special wrench indispensable when fitting or removing taps on the bath, and also certain types of basin taps. Sometimes called a crowsfoot spanner, its jaws fit around the retaining gland nut or compression nut with the axis of the spanner running down the feed pipe. You can get leverage on the spanner by putting a steel bar through the lower jaws.

Tank cutter or hole saw You will want this type of cutter if, for example, you need to plumb in pipe to the cold water storage cistern. A range of sizes is available, depending on the diameter of the pipe you are fitting. The cutter comes as an attachment to an electric drill. Always operate the drill at a slow speed.

An alternative tool for this job is the chassis punch. With this, you first drill a relatively small hole in the side of the cistern. Hold the punch on one side over the hole and the die on the other. Feed the screw provided through the punch and the pre-drilled hole into the die. As you tighten the screw, the punch will cut through the side of the cistern. Various sizes are available depending on the pipe being fitted.

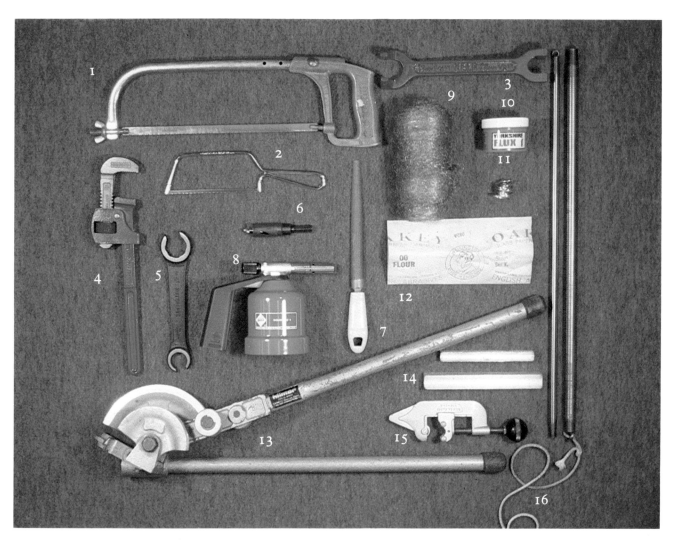

This range of tools will be useful for installation work: 1 Large hacksaw, 2 Junior hacksaw, 3 Basin and bath wrench (crowsfoot spanner), 4 Stillson wrench, 5 Compression joint spanner, 6 Hole saw, 7 Half-round file, 8 Butane blowtorch, 9 Steel wool, 10 Flux, 11 Solder, 12 Glasspaper, 13 Bending machine, 14 Bending machine formers, 15 Pipe cutter, 16 Bending springs.

3 Basic maintenance

Problems always have a habit of coming at the most inconvenient time. In the case of the plumbing system, you not only face the aggravation of having to wait until a plumber can call, but also the expense incurred for a job that you could just as easily have done yourself. What would cost you sometimes a few pence for materials can end up as pounds when you get someone in.

The following are routine maintenance and repair jobs the competent DIY enthusiast can tackle – and it is particularly useful to be able to do this kind of work if you are faced with an emergency.

Repairing dripping taps

This is probably the most common fault in the domestic plumbing system. Apart from the inconvenience and the staining that it can cause on certain types of basin and sink, a much more serious problem can result if you do not repair the tap. In extreme cold conditions, there is a far greater chance of water freezing in this situation, particularly if the tap in question is outside or in an exposed, unheated area.

With new technology, a type of tap is being introduced that does away with the traditional washer, which, when worn, causes the tap to drip. With this latest design the flow of water is controlled by two ceramic discs which do not wear. But the old style of tap will be with us for many years yet and it is important to know how to change the washer as soon as it shows signs of wear. The instructions given here apply to the conventional bib and shrouded-head taps and the Supatap.

Taps with ceramic discs held in a cartridge do away with the need for standard washers.

Check on the type of tap washer you need from the range available.

If you notice the dripping straight away, it may be that the washer has not yet worn badly, but that a foreign body has got wedged between the seat of the tap and the washer. Try turning the tap on full two or three times. If the dripping stops, then the fast flow of water will have washed away the foreign body. If not, then you will have to replace the washer as follows:
1. *Buy the correct size washer.* The size of a tap is described by the size of the pipe feeding it. Bath taps are therefore usually 22mm (or ¾in), while normally the rest of the taps in the home will be 15mm (or ½in). Traditional leather washers have been superseded by synthetic

rubber ones, normally flat discs with a hole in the middle. Hemispherical washers are available and these will give better service in a tap with a worn seating.

2. *Have the right tools ready.* Depending on the type of tap you are working on, you will need a spanner to remove the body from the tap and, if the tap has a chromium-plated cover over the body, one to fit that as well. Remember to wrap a cloth round any chromium-plated parts of a tap when using a spanner to prevent the jaws damaging the finish. You will also need a small spanner to remove the nut from the jumper inside the tap and a screwdriver to prise off the old washer.

3. *Turn off the water.* If the tap is connected direct to the main water supply (which means it works at high pressure), turn off the water at the main stopcock, usually located under or near the kitchen sink. The mains-fed taps will usually be the cold water one at the sink and any outdoor ones.

If the tap is supplied from the cold water storage cistern in the loft or from the hot water cylinder, which means it works at low pressure, turn off the gate valve on the feed from the cistern or the cylinder. If no such valve is fitted, you will have to turn off the main stopcock as before and drain the system (see page 36). If you have to drain the system, make sure you have turned off the water heating system, otherwise you could cause damage to the boiler or calorifier in the hot water cylinder.

Finally turn the faulty tap on fully to make sure there is no more water left in the feed to it. (For Supatap, see 9.)

4. *Unscrew the cover or shroud.* With the older type of tap, you will first have to unscrew the chromium-plated cover, which is usually not that easy. If you have trouble moving it, pour on hot water to help release it. If there are no flats round the cover on which a spanner can grip, wrap a large thick elastic band round to provide a better grip. Failing this, wind a strip of emery cloth two or three thicknesses deep round the cover and, with a self-grip wrench, carefully unscrew it.

With this type of tap you do not have to remove the handle when replacing a washer. With plastic shrouded-head taps, the handle usually just pulls off. Occasionally there may be a small grub screw at the side to hold the handle in place or a screw under the hot or cold label on the top. In this case prise off the label carefully with a knife or small screwdriver and undo the screw.

5. *Unscrew the body.* Make sure you have the right size spanner to fit snugly over the lower hexagon of the tap body and unscrew this carefully to release the headgear from the rest of the tap. If you have difficulty moving the headgear, wrap some padding round the base and grip this with a suitable size spanner while you turn the body. This will prevent the whole tap twisting in the wash-basin, bath or sink.

6. *Inspect the seat.* Look down into the bottom of the tap and check the seat. This should be clean and free of scratches or grooves. If it is damaged, you can buy a nylon washer and seating set to effect a repair. The kit comes with full fitting instructions. This method is much easier and cheaper than re-cutting the seat.

7. *Replace the washer.* The jumper, which contains the washer, will either be still in the body of the tap you have removed or lying over the seat in the base of the tap. The washer is held in place with a small retaining nut, which you must unscrew. A little penetrating oil may help to release this nut if it is tight. Prise off the

The shrouded-head tap handle either pulls off or is released by a screw underneath the label.

The rising spindle type of tap contains a packing-filled gland in the headgear.

Release the headgear by undoing the hexagonal nut with a spanner or good adjustable spanner.

When you remove the headgear, you can take out the jumper and replace the washer.

old washer, replace it with a new one and fix this in place by tightening the retaining nut.

8. *Re-assemble the tap.* Put the jumper and washer assembly back in the seat of the tap, then smear a little petroleum jelly on the screw threads. Check that the tap is turned full on, so that the jumper is as near as possible to the body. Screw the body back onto the base carefully, making sure that the threads are not crossed, tighten up the body and replace the cover or shrouded head, depending on the type of tap.

Turn the tap off before you restore the water supply, then open the repaired tap slowly to get rid of any trapped air first. When you turn the tap off after use, twist it until the drips stop – and no further. This will prevent undue wear on the washer in future.

9. *Replacing a Supatap washer.* The great advantage of this type of tap is that you do not have to turn off the water supply while you replace the washer. Turn on the tap roughly one revolution and, with a suitable size spanner, release the nut at the top of the handle. Turn on the tap again and keep turning. The water flow will reach its maximum and then stop, when the tap nozzle will come off in your hand.

If you turn the nozzle upside down, the anti-

To repair a Supatap, unscrew the hexagonal tap top anti-clockwise to free the mechanism.

Turn the tap clockwise until the water flow stops, then turn again to remove the nozzle.

Tap the base of the nozzle on a hard surface to release the washer and jumper and replace this assembly. Use the reverse procedure to put the tap back together again after the repairs are done.

splash device with the washer and jumper will fall out. You can free the jumper and washer assembly by inserting the blade of a screwdriver between it and the anti-splash device. With the Supatap, you have to replace the whole assembly.

Re-assemble the tap, bearing in mind that when you screw back the nozzle to the fixed part of the tap it has a left-hand thread.

Repairing leaking taps

Sometimes a fault develops in the tap when the packing in the spindle gland wears down. You will notice this when the tap is running, because water leaks out round the handle. Another effect of worn packing is that the handle turns very freely. Fortunately you do not need to turn off the water supply to solve the problem.

With the traditional bib tap, you must unscrew the cover as described earlier when changing the washer (see page 25). Then remove the grub screw that holds the handle in place and take off the handle. If this is too stiff, turn the tap full on and lift the cover until it is against the underside of the handle. Place two suitably sized pieces of hardwood between the bottom of the cover and the base of the tap and start turning off the tap. The pressure applied will push the handle up the spindle and you can then remove it and the cover. With a shrouded-head tap, take off the head as earlier described (see page 25).

You will now be able to see the gland nut, which is a small hexagon at the top of the tap body. Tighten this nut half a turn and try the tap again. If it still leaks, keep turning the gland nut until the leak stops. If you find there is no more adjustment available and the tap still leaks, turn it off and unscrew the gland nut.

Remove the old packing with a small screwdriver and fill the gland to within 3mm ($\frac{1}{8}$in) from the top with a proprietary packing material, tallow cotton or greased string, winding it around the tap spindle in a clockwise direction. Refit the gland nut and tighten it gently; the spindle should feel slightly stiff.

Finally, re-assemble the tap. If you find the handle difficult to replace onto the spindle, check that the grub screw holes line up and then tap the handle gently using a piece of softwood as padding between the handle and the head of the hammer. Then replace the grub screw.

If you cannot see a hexagonal gland nut when you take off the cover, it means your tap has an 'O' ring gland. First buy a new 'O' ring and then turn off the water supply. Remove the headgear as you would when changing a washer (see page 25) and unscrew the spindle right out of the headgear. You will now see the 'O' ring in the hole from where you took out the spindle. Prise out the old ring with a small screwdriver and insert the new one. Lubricate it with a little petroleum jelly and then re-assemble the headgear.

Tighten the gland nut to stop the spindle leak; if not, you must replace the gland packing.

Remove the old packing and wind tallow cotton in a clockwise direction into the gland.

If, when stripping down the top of a bib tap, you find the handle is too stiff to remove, use two wood blocks under the cover and start turning the tap off to free the handle.

If there is no gland nut in the tap, then it has an 'O' ring seal. You will have to remove the headgear from the tap and take out the spindle if you need to replace this ring.

Repairing a cistern valve

A sure sign that you have got problems with a cistern valve, whether it be the cold water storage cistern in the loft or one of the WC cisterns around the house, is when water drips or runs out of the overflow pipe.

If you check on the outside of the house, you will see these short lengths of pipe sticking out from the wall. If you see water coming out from any of these, you must rectify the fault in the cistern as soon as possible. If you do not tackle the job straight away, not only will you waste water, but you run the risk of the water overflowing from the cistern. This can prove very expensive if water damages decorations, furniture or fittings.

The danger is most acute in frosty weather, since the overflow pipe will very quickly freeze up and cause the cistern to overflow that much sooner.

You can check where the fault lies by removing the cover from the cistern and gently lifting the float arm inside. This arm is designed to operate the valve according to the level of water in the cistern. When the water reaches a predetermined level the valve should shut and cut off the supply of water into the cistern.

If, when you lift the float arm, the water stops flowing through the valve you know the fault lies with the arm or the float. If water continues to flow, then the fault is in the valve itself.

Repairing the float or arm The following procedures are for brass valves only. Turn off the water at the main stopcock, if working on the cold water storage cistern, or the gate valve controlling the flow of water from the storage cistern. Then unscrew and remove the float.

Shake it to see if it contains any water. If it does, you will have to replace it since it has obviously developed a leak. If not, replace the float and bend the arm down slightly. Turn the water back on and check whether the valve cuts off when the water reaches the correct level – usually about 25mm (1in) below the overflow pipe. Continue bending the float arm until the water stops running.

There is obviously a limit to the amount of adjustment you should make to the arm. If you have to adjust it much more than 25mm (1in), then you should check the valve itself.

Repairing the valve There are three basic types of cistern valve – the Croydon, the Portsmouth and the more modern Garston. The Croydon is now virtually obsolete. If you do have one of these still fitted in the cistern, you would be best advised to replace it with the Garston type, which will give much better and quieter service.

Portsmouth valve This is probably still the most common with existing systems. You can repair it provided that the valve seating is not damaged. To check this valve, turn off the water supply at the stopcock, close up the ends of the split pin holding the float arm in place, remove the pin and lift the arm clear.

Check whether there is a screwed cap on the inner end of the valve and, if there is, remove it using pliers if necessary. You can now remove the valve piston, which you can ease out with the help of a screwdriver blade in the slot underneath.

Take a look inside the valve and check on the valve seating. This should be clean and form a complete circle without any pit marks or scratches on it. If you see any damage, you will have to replace the whole valve. In this case, make sure you replace it with one of the correct pressure rating, since valves are available for either high or low pressure use. High pressure valves are used to control water direct from the mains supply, while low pressure ones are fitted where the supply is from the cold water storage cistern. The high pressure valves have a smaller seat diameter.

Provided the seating is undamaged, look at the end of the piston where the valve seating fits. Inside you will see a flat rubber disc – and this is almost certainly where the problem lies.

To replace this rubber washer, you have to unscrew the cap end of the piston from the main body. Grip the cap end firmly with a pair of pliers and release it. It will help if you grip the body of the piston at the same time, either with another pair of pliers or by inserting the blade of a screwdriver into the slot to prevent it rotating.

Depending on how long it has been since the cap was last removed, it may be very stiff and often the joint is difficult to see.

Once you have unscrewed the cap you can take out and replace the old washer. Smear the

Using pliers, close the open end of the split pin under the valve and pull it out carefully.

With pliers or a self-grip wrench, release the valve cap by turning it anti-clockwise.

The Portsmouth valve is the one most commonly found in both the cold water storage cistern and the WC cistern, although it is now being replaced by the Garston valve.

Having taken out and replaced the rubber washer in the valve cap, screw the cap back on, remove any burr from the piston and fit this back into the valve. Replace the arm.

screw threads with a little petroleum jelly before re-assembling the piston. Clean up the piston before you put it back in the valve body, using fine glasspaper to remove any burrs caused by the pliers gripping the outside. Wipe the piston thoroughly with a damp cloth to remove any dirt and smear it all over with petroleum jelly.

Slide the piston back into the valve body and refit the float arm, not forgetting to bend back the ends of the split pin. Replace the screwed cap, where fitted, and turn the water back on. Allow the valve time to shut itself off as the water level rises. If the level is too low, then bend the arm up slightly. If it is too high, you can lower the arm a bit.

Some Portsmouth valves are fitted with a silencer tube that runs down below the water level. Most water authorities have now banned the use of these tubes, since there is a danger of water being siphoned back from the cistern into the mains in the event of mains pressure failure. You can prevent this happening by drilling a small hole of about 3mm ($\frac{1}{8}$in) diameter through the top of the tube just below the valve.

Garston valve This is the most modern type and is made of either brass or plastic. It has a tough, score-resistant nylon seating and a large rubber diaphragm which sits against it when the valve is closed.

When used in a cold water storage cistern, the valve is fitted in the normal way to the side of the cistern. On some modern WC cisterns, however, the water feed enters from the bottom and in this case the valve is fitted at the top of a plastic tower, which rises from the base of the cistern.

The delivery of water may be from below the valve through a silencer tube. More recent installations, however, incorporate an anti-siphon delivery tube above the valve. This directs water against the side of the cistern to reduce noise and prevent the risk of water siphoning back.

The Garston valve has a replaceable seating, which is available for use with both high and low pressure supply.

You can make adjustments to this type of valve in the following way: turn off the water supply to the cistern and flush the WC fully to allow the ball float to drop. At the opposite side of the valve to the ball float you will find a brass steadying screw, the head of which should just touch the side of the cistern. If it is not, loosen the lock nut and turn this screw until it does touch the side. Then tighten the lock nut to hold the screw in place. Turn the water supply on and watch to see if this has cured the problem.

If this does not work, lift the float valve gently to see whether it shuts off the water supply. If it does, this means you will have to adjust the float arm. The difference with the Garston valve is that you must not try to bend the arm.

You will see the float arm is fixed to a pivot

The modern Garston valve is easier to adjust and work on than the traditional types. There are two kinds – one that is side-fed and one that is fed from the bottom of the cistern.

With the Garston valve, adjust the steadying screw so the end just touches the cistern.

Move the float arm away from the valve to raise the water level and nearer to lower it.

To remove the diaphragm, unscrew the large hand nut. You can pour hot water over to release it.

Check the seat is sound, lubricate the screw threads and fit in the new diaphragm.

bracket with two lock nuts, one each side of the bracket. Turn off the water supply to the cistern, flush the WC and then slacken the lock nut furthest from the valve half a turn. Tighten the other lock nut to take up the slack, turn on the water and allow the cistern to fill up before checking the level again. Repeat this procedure as needed until you get the right level of water.

If these adjustments do not stop the flow of water into the cistern, you will have to replace the valve diaphragm.

Turn off the water supply to the cistern and undo the large serrated-edged nut nearest the float. This will probably be quite tight, but you should be able to free it by pouring hot water over it. If necessary, use a self-grip wrench, but make sure you are also gripping the body of the valve with another wrench or adjustable spanner. You can now remove the diaphragm.

Check the plastic seating against which the diaphragm presses. If it is damaged at all, you can replace it by unscrewing the other large nut on the body of the valve. Make sure the new plastic seating is the same size. Fit in the new diaphragm and smear a little petroleum jelly on the screw threads before replacing the nuts. When tightening plastic nuts, always use hand pressure only.

Torbeck valve This is a more sophisticated version of the Garston valve and may be fitted to overcome fluctuations in water pressure or water hammer caused in other valves by the float arm bouncing against the high pressure of incoming water.

The valve is easily identifiable, since it has a very small plastic float on a short arm. It operates in a similar way to the Garston valve but is so designed that water pressure is acting equally on both sides of the valve diaphragm. The valve is made for use with high pressure feed and does not have an interchangeable nozzle.

To change the diaphragm involves the same techniques as for the Garston valve, but using, of course, the special Torbeck fitting. This has a small hole in it which must be fitted over the pointed metering pin contained within the body of the valve.

To adjust the water level in the cistern, you can clip the small plastic float arm in different positions on its bracket. By raising the float arm you raise the water level and vice versa.

Replacing a WC cistern diaphragm

Another problem with the WC cistern is when you get diaphragm failure. You will notice this when the flushing mechanism becomes harder to operate and often you will have to try several times to get the cistern to flush.

First remove the lid and check the water level since it could be that there is just not enough water in the cistern. Remember the water should come to about 25mm (1in) below the overflow pipe. To increase the level, either bend the float arm up carefully or – with a Garston or Torbeck valve – make the necessary adjustment.

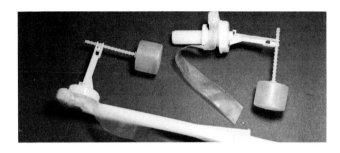

The Torbeck valve is designed to overcome fluctuations in water pressure.

Water pressure acts equally on both sides of the Torbeck valve's diaphragm.

If this fails to solve the problem, you will have to remove the siphon unit and replace the diaphragm. First you must turn off the water supply to the cistern and flush the WC fully to empty it. If the diaphragm failure is that severe and the flush hardly operates, then you will have to bale out the water with a small jug or mug. Mop up the rest of the water in the bottom of the cistern with an absorbent cloth or rag.

Cisterns of course do vary in design. It may, for example, be necessary to remove the complete ball valve before you can take out the siphon unit. If yours is a modern close-coupled type, where the cistern sits on an extension of the pan, you will have to remove the cistern.

Disconnect the water supply pipe where it enters the base of the cistern by unscrewing the union nut that holds it in place. Use the same method to disconnect the overflow pipe on the other side. You should find the cistern is screwed to the wall from the inside. Undo and remove these screws. Finally locate the two wing nuts at the back of the WC pan on the underside of the cistern and unscrew them to release the cistern.

With the traditional type of cistern, which is fixed to the wall above the WC pan, all you need do is disconnect the downpipe by unscrewing the large nut underneath the cistern. Make sure you have a bowl handy to catch any water left in the cistern.

Disconnect the operating link with the handle, unscrew the large back nut beneath the cistern and withdraw the siphon mechanism. You will see the siphon plunger at the bottom of the siphon tube. Remove the plunger to expose the diaphragm, which is on the top of the plunger, and take off the old diaphragm.

When you fit the new one, you must make sure it is the correct size. Insert it into the siphon tube and test that it just touches the sides, but does not drag. Always get a diaphragm big enough or slightly larger, in which case you can carefully trim it down to size with a pair of sharp scissors, using the plate at the end of the plunger as a guide.

Re-assemble the siphon unit and replace it in the cistern, connecting up the operating link with the handle. When you connect the downpipe, use non-setting mastic around the large nut at the base of the cistern. If working on the self-contained unit, make sure you replace the gasket between the pan and the cistern.

Reconnect the overflow and water supply pipes and turn the water back on. Check that none of the joints are leaking. The cistern should be ready for use in a couple of minutes. If it takes longer than this, check the nozzle in the valve is a low-pressure one (with a large hole), unless the cistern is connected direct to the main supply.

Changing from single to dual flushing This refinement, which has been introduced on most modern cisterns, enables you to operate half-flush if the handle is pressed and then released and full-flush if the handle is held down. By using the half-flush, you can save considerable amounts of water over a period – and this is particularly useful in times of a water shortage.

At the top of the large diameter of the siphon tube you will see a hole into which a plug is fitted. The plug supplied is a dual one, with one

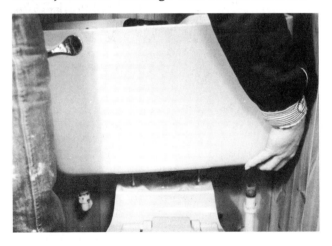

To remove a close-coupled cistern, turn off the water, flush the cistern, disconnect the inlet and overflow, take out the back retaining screws and undo the bottom wing nuts. Then lift off.

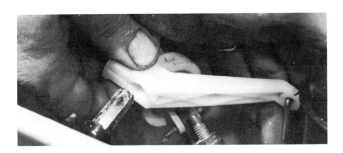

Disconnect the lever linkage and remove the float arm if needed to clear the siphon.

Unscrew the large nut to remove the siphon. Make a note of the position of the washers.

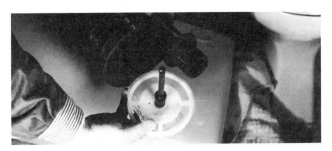

Trim the new diaphragm to fit closely inside the siphon without dragging on the sides.

Fit the plug with a hole for double flushing and the solid one for single flushing.

section solid and the other with a hole in it. If you fit the plug with the hole, you can operate the dual flushing system. With the solid plug, you will get just single flushing.

Freeing a jammed stopcock

Because in the average home little work is normally done on the plumbing system, stopcocks and gate valves – which control various parts of the system – are rarely operated. And because of this lack of use, they can jam open.

With valves that control individual parts of the system, this will be inconvenient if you want to isolate the water supply to individual fittings. But with the main stopcock, which controls the supply of water into the home, this can be disastrous if you have an emergency on your hands.

It is a warning often given, but equally often ignored, but you should check the main stopcock – and ideally all the other control valves in the system – once a month to ensure they do not jam. Turn them off and on two or three times to check they are not sticking.

In an emergency, if the main stopcock is jammed, you may be able to turn off your supply at the water authority's stopcock outside the house. Unfortunately, however, this sometimes has a shaped shank that can only be turned with a special key. So it is most important your main stopcock is working.

If the stopcock is jammed, you should be able to free it by cleaning round the gland with a wire brush and squirting a little penetrating oil in as well. Leave the oil to work in for about 15 minutes and then try turning the handle. Do not forget that this should be turned in a clockwise direction when trying to shut off the stopcock.

If, after several applications, the stopcock still will not turn, you can try moving the handle by packing either side with small pieces of softwood and tapping gently with a hammer. Avoid using excessive force here, since the handle can snap quite easily.

If, after trying these methods, you cannot turn the stopcock, you will have to take it apart. This involves turning off the water at the authority's stopcock – and you may need to obtain a special key to do this. You dismantle the stopcock in the same way as a bib tap (see page 25). Remove the washer and jumper, unscrew the gland and take out the packing. This will make it easier to work on the body and spindle of the tap, which you should clean up, free off and lubricate with penetrating oil or paraffin. When you have eased off the handle, re-assemble the stopcock. Make sure you grease the screw threads with petroleum jelly and repack the gland (see page 28).

Draining the system

You may find it necessary to drain the water from the hot or cold water systems in the home, for example to replace the cold water storage cistern or carry out repairs to the boiler. Before you do this, there are two precautions you must take. The first is to turn off the main stopcock to prevent the system from filling up again. The other is to switch off all forms of water heating to prevent damage to the boiler, immersion heater or calorifier (see pages 14–17).

Both the primary and secondary water systems will have a draincock fitted for this purpose – and these will be situated at the lowest level in the particular system. The draincock on the rising main is usually next to the main stopcock. On a primary system it is near the boiler on the return pipe from the storage cylinder, while on a secondary system it is usually below the kitchen sink hot tap or the lowest hot water fitting on the circuit.

To drain off the water, fit one end of a length of hosepipe over the fitting on the draincock, securing it in place with a hose clip. Make sure the other end of the hosepipe is at a lower level than the draincock and avoid raising the hosepipe along its length as much as possible. You will always have to feed the hosepipe outside, preferably away from the house or into a drain.

Use a suitable size spanner to unscrew the draincock until there is a good flow of water through the hosepipe. Take care when unscrewing the draincock, since with some of the cheaper ones the spindle can be unscrewed right out of the body of the tap – and the consequences can be disastrous if the system empties all over the floor. Remember to tighten up the draincock afterwards.

Repairing leaking joints

If it has been made up correctly, a joint in the plumbing system should not leak unless it has been disturbed or damaged by frost. This is why

Attach a suitable length of ordinary hosepipe with a clip to remove water from the system via a draincock. Do not loosen the draincock so much the spindle falls out.

you should always be careful when working on sections of the system that you do not exert too much pressure on the pipes, since this can cause a chain reaction in the adjoining joints.

If a capillary joint starts to leak, the only way to cure it is to remake the joint (see pages 48–50). If it is a compression fitting that is leaking (see pages 50–52), the first thing to do is to try tightening the nut slightly, while keeping the fitting in place with another spanner. This may stop the leak. If it does not, turn off the water supply at the nearest control valve and, with a suitable bowl or bucket underneath, slacken the nut at whichever end the joint is leaking. Dry off the joint with a clean cloth, wrap two or three turns of PTFE jointing tape around the compression ring and tighten up the nut again.

If this does not stop the leak, you will have

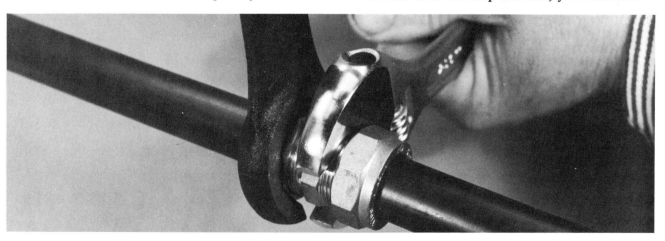

If you have a leaking compression joint, you may be able to stop the problem by tightening the joint. But to do this you will need two spanners, one to hold the fitting in place and the other to tighten the nut.

If this does not solve the problem and the leak continues, you will have to take the compression joint apart and wrap some PTFE jointing tape around the compression ring.

to remake the whole joint using a new compression ring on a new piece of copper pipe. If the joint is at the end of a long section of pipe, cut a suitable length piece off this pipe, rather than remove the whole length. In this case, you can fit the new piece of pipe to the cut end of the original pipe using a straight connector.

Normally plastic joints should not leak. If they do, however, you will have to remake the joint (see pages 60–65).

Lagging pipes

All water pipes in exposed and unheated areas of the home should be lagged to prevent them from freezing up in extreme cold weather. The alternative – a burst pipe – is not only inconvenient but can be very expensive, depending on the extent of the damage caused by lost water.

In the interests of economy you should also lag all your hot water pipes, except of course those in an airing cupboard. Heat lost in the pipe runs is fuel wasted – and there is little point in allowing this inefficient form of heating to warm up the loft area or underneath the floorboards.

There are several good insulating materials available to wrap round your pipes. One of the most convenient and easy to fit is plastic foam, which comes ready split down its length and will fit snugly round the pipes.

Make sure you buy the correct size, since the lengths are available in different diameters to fit individual sizes of pipe. You can get this lagging with a continuous plastic zip fastener down the split – and the material is sufficiently flexible to

You will find a range of lagging materials to protect the piping in your domestic plumbing system. This includes (from left) plastic 'zip-up' foam insulation, 15mm split foam insulation that is held with PVC tape and wrap-round glass fibre bandage. This is also held in place with PVC tape and is especially useful for awkward areas, particularly round taps and valves where foam can be difficult to use.

cope with bends and capillary fittings.

When you lag the pipes, do not forget to do the awkward areas – such as round taps, valves, tees and elbows – as well. For these areas you can use odd bits of the lagging material and hold them in place with PVC tape. When you work round a tap or valve, make sure you can still turn the handle on and off when needed.

You will have to fit a waterproof, non-absorbent material over the lagging when treating outdoor pipes to prevent the lagging from freezing. If you use the plastic foam lagging and cover it completely with black polythene held in place with PVC tape, this should last for about five years before the polythene breaks down. If you want to bury water pipes to keep them away from the frost, you will have to go down about 750mm (30in) to ensure complete protection.

Other materials are available for lagging pipes; these include moulded polystyrene and glass fibre bandage but generally they are not so convenient to fit. You can use expanded polystyrene chippings or vermiculite (which are used for loft insulation), but with these you will have to enclose the pipework in a box and fill this with the chippings. The old practice of using newspaper bound with rags is nowhere near as effective as any of the materials mentioned.

Lagging a cistern or cylinder

You should certainly lag your cold water storage cistern and this is best achieved using flat sheets of expanded polystyrene or insulating board. But when you fit these, remember not to insulate the bottom of the cistern, since warm air from the rooms below will help prevent the water freezing. This means that no loft insulation

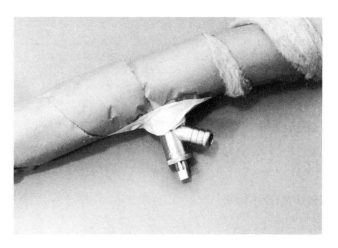

When you lag pipes, do not forget to insulate round the valves as well. Use PVC tape to fix the lagging, but leave the handle free.

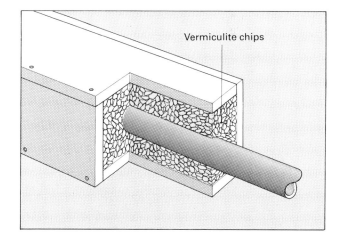

An efficient way of lagging pipes in exposed situations is by boxing off that section of piping and filling the box with vermiculite chips. This would be ideal on outside piping.

should be laid in this area either. You can use PVC tape to hold the sheets in position.

You must also cover the top of the cistern, but remember here to leave a hole to take the 'U' bend at the top of the hot water cylinder vent pipe.

Insulating jackets are available to lag hot water cylinders and straps are provided to tie the jackets in place. You will still get some warmth escaping into the airing cupboard.

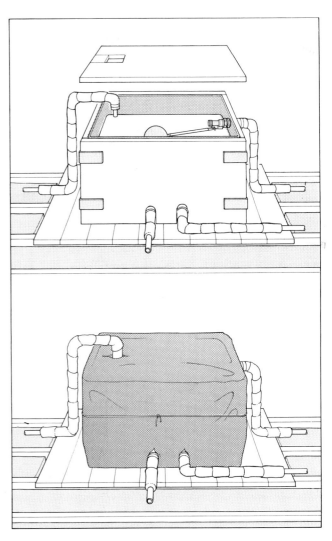

Depending on the type of cold water storage cistern you have in the loft, you can either fit a proprietary jacket or fix on sheets of polystyrene or insulating board with PVC tape.

With the modern circular plastic cistern, you can buy a special insulating jacket to fit. Usually all you need do is tie the jacket in place round the cistern.

You should also lag the hot water cylinder with a suitable insulating jacket to prevent heat being wasted. This is still necessary even if the cylinder is situated in an airing cupboard. Sufficient heat will still escape to cope with the airing of freshly washed clothes and other laundry.

Repairing burst pipes

You will soon know if any of your pipes are frozen, because water will not flow through the relevant tap or valve. At this stage there will not be any evidence of a burst pipe since the water inside will still be frozen. Unfortunately when water freezes it expands with considerable force. This can either burst the pipe or dislodge a fitting at the end of the frozen section.

Look for the frozen section immediately to check what damage has been done and to prevent further damage being caused. The obvious places to look are in exposed or very cold areas – for example in the loft – or in draughty places or where the sun never reaches. Inspect the fittings at either end of these sections to see if they have been moved at all.

When you have located the frozen section, turn off the water supply at the nearest available stopcock and turn on the affected tap. Then warm the pipe gently to melt the ice. Ideally you should use hot wet cloths or hot water bottles. If you use a hair dryer or fan heater, make sure you keep the appliance away from the pipe in case there is a split. On no account must you get water on the appliance.

Heat the whole area of pipe until the ice

When repairing a split section, cut out the damaged pipe and clean up the cut ends.

Slide on the end of the repair pipe without a pipe stop then fit the other end up to the stop.

Tighten one nut, then hold it firmly with a spanner while tightening the other nut. Make sure when you tighten the fitting that the pipe is not being twisted. Lag the whole section of pipe after completing the repair.

thaws and starts flowing out of the tap. Keep a large bowl or bucket handy in case you can catch the worst of the spray if the pipe has been split.

Of course, a blowtorch could be used to heat up the pipe, but this can be very dangerous since the pipe will almost certainly be running next to floorboards or against a wall and you do not want to start a fire as well. So use a blowtorch only as a last resort – and then with extreme care.

If the pipe has burst, in which case water will either drip or spray out, make sure you drain that section until there is no more water in the pipe. Then examine the pipe to see how extensive the damage is. If the split is a small one, you can use a frost repair kit.

These kits are suitable for repairing small splits – up to 87mm ($3\frac{1}{2}$in) in copper or stainless steel pipe. The repair fitting is a straight compression coupling with only one pipe stop.

To fit this coupling, first turn off the water and drain the pipe (see page 36). Cut out the split section, leaving a gap in the pipe of no more than 87mm ($3\frac{1}{2}$in), and remove the burr from the cut ends.

Slacken the compression nuts on the coupling and spring one end of pipe clear so you can slide the coupling onto it. Make sure the end of the coupling you slide on is the one without the pipe stop. Realign the pipes and slide the other end of the coupling onto the other end of pipe up to the pipe stop. Tighten both the compression nuts by hand and then another two-thirds of a turn with a spanner.

If the split section is longer than 87mm ($3\frac{1}{2}$in), you will have to fit a new piece of pipe of a suitable length. If you are fitting a short length of pipe into the existing system, you can use two slip couplings, fitting them as above.

If the damage is extensive or at a bend in the pipe, you will have to replace that section with a new piece of pipe (see pages 37–38). If the fitting at the end of the affected section is leaking, you will have to remake the joint (see pages 50–52).

You can make a quick temporary repair with an epoxy resin repair kit, if the damage is not too extensive. There are now several types available, including one that claims to be suitable for use on wet pipes. As a general rule, however, you should always make sure the area around the burst is dry before you start the repair.

Hammer the split closed with a wooden mallet and clean around it with glasspaper. Apply the resin over the split and around the pipe according to the manufacturer's instructions. While the filler is still tacky, wrap

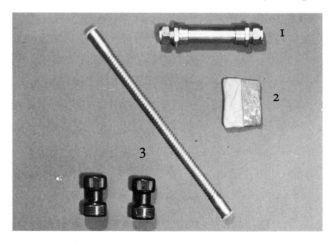

Among the range of frost repair kits most commonly available for burst pipes are: 1 Slip coupling for short splits, 2 Epoxy resin, 3 Kopex flexible pipe for longer splits.

glass fibre bandage firmly around it and finish off with another layer of resin. Leave it to dry hard and then clean off the repair with glasspaper.

Manufacturers do claim that these kits form a permanent repair, but it will look unsightly and you may decide you want to make a permanent repair later if that section of pipe is exposed to view.

Once you have repaired the burst, do not forget to lag the pipe to prevent the problem recurring.

Emergency action The first indication that a pipe is frozen will normally be when water fails to run from a tap or through a cistern valve. However you may not discover that you have a frozen pipe until the ice starts melting and leaks from a split or damaged joint.

As water freezes, it expands and causes the pipe to stretch. Sometimes the pipe can accommodate this expansion, but more often than not the pipe will split or push the joint unions along the pipe.

If you see patches or drips of water – down a wall or on the ceiling, for example – turn off the water supply at the main stopcock immediately. Then go into the loft and turn off the two gate valves at the bottom of the cold water storage cistern. Next turn off the boiler and central heating pump, if fitted and working.

You must also check that no electrical fittings or circuits are affected by the leaking water, although you may not discover this until you have located the damage. If this is the case, turn off the electricity at the main switch on the consumer unit immediately. Water and electricity are a lethal combination!

Now find out where the damaged pipework is. Bear in mind when you are looking that water will run down a vertical or sloping pipe before it drips off and in this situation the leak may well be much higher up the pipe than the position of the drips would suggest. The first places to look are around the visible damp patches or where the pipework is exposed, for example down exterior walls or in the loft space.

Clearing a blocked pipe

The most common type of blockage is at the kitchen sink and this is normally caused by fat setting in the trap beneath and thus blocking the waste. Hairs in the wash-basin or bath are another culprit.

First try pouring a kettle full of boiling water down the plug hole. This may clear the blockage and, even if it does not, it will help free the drain plug if the U-bend underneath is made of metal. The thread of these drain plugs has a

When using a plunger to clear a blocked pipe, stop up the overflow outlet and, with a little water in the bottom of the fitting, give the plunger several sharp pumps. This should clear simple blockages.

habit of seizing up and thus becoming very difficult to turn. You can also try pushing a length of old spring curtain wire down the waste pipe in case you can shift the blockage this way. In the case of hairs, this should do the trick.

If you have so far failed to shift the blockage, try working it free with a suitable sized plunger. Run a little water into the sink, wash-basin or bath (if there is none there already) and push a damp cloth firmly into the overflow. Give the plunger several sharp pumps.

If this does not work either and the trap is metal, you will have to remove the drain plug. Put a container under the waste trap and, using a suitable tool such as a small rod or bar, unscrew the drain plug. Remember to turn it anti-clockwise (as you look up at the plug). Clear out the blockage, replace the plug and flush the waste pipe with plenty of hot water.

If you cannot remove the drain plug, see if caustic soda will clear the blockage. This is a very quick-acting chemical that generates heat when mixed with water. Since the reaction is quite violent, you must use this substance with great care. Wear strong rubber gloves when handling it and follow the manufacturer's instructions thoroughly, particularly on the action to take if you get any caustic soda on your body.

Failing these methods, in the final resort you will have to take out the trap. To do this, unscrew the union nuts at either end with a large adjustable spanner. Clean the trap out thoroughly and fit it back in position.

The easiest type to clean is the plastic bottle trap. You simply unscrew the large hand nut at the bottom of the trap. Make sure you put a bowl or bucket underneath to catch the contents of the trap.

With a 'U' shaped trap, you can remove the

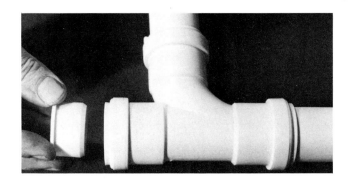

You can get some fittings that incorporate access for a rod to clear blockages, such as the cleaning access provided on this swept tee.

first half of the bend by unscrewing the two large plastic nuts; one holds the trap to the sink and the other is situated at the bottom of the bend. This will give you access to both parts of the trap, which you can then clean out.

Before you put the trap back, smear a little petroleum jelly on the thread to ensure the trap can be screwed back in position easily and tightly.

It may be that the blockage is in the waste pipe and not the trap. In this case the method you use will depend on the system you have. With metal pipe you should be able to dislodge the blockage with a drain rod or flexible drain cleaner. There may be a removable end plug on a bend or elbow fitting which will enable you to use a straight rod through the pipe.

You will find with the original lead pipes that a build-up of slime is the main culprit when a pipe gets blocked and here a caustic soda-based cleaner should remove the problem. As already mentioned, always follow the manufacturer's instructions carefully when using this type of cleaner.

4 Metal components

Traditionally plumbing was always in metal and today a full range of components and fittings are still available. Although plastic fittings are gaining in popularity, the majority of domestic systems is still in metal.

Copper pipe

Copper is the most popular type of piping used in domestic plumbing and is made to BS 2871 Part 1. It normally comes in long lengths, although most suppliers will cut you minimum lengths of a metre.

The two most common sizes are 15mm and 22mm, which roughly correspond to the old ½in and ¾in sizes. Bear in mind, however, that metric and Imperial sizes are not completely interchangeable.

If you are using capillary fittings, you will need special adaptor couplings if you are joining new pipe to existing Imperial size pipe.

With compression fittings, you can join 15mm to ½in pipe, but you will need an adaptor connector with 22m and ¾in pipe.

Stainless steel pipe

This type of pipe is available in 15mm and 22mm sizes to BS 4127 Part 2, but is very expensive and often difficult to obtain. You will need a bending machine to shape it and can only use it with compression fittings, not capillary ones, since you cannot solder stainless steel. If you do use it, you will have to tighten the fittings more than with copper pipe because

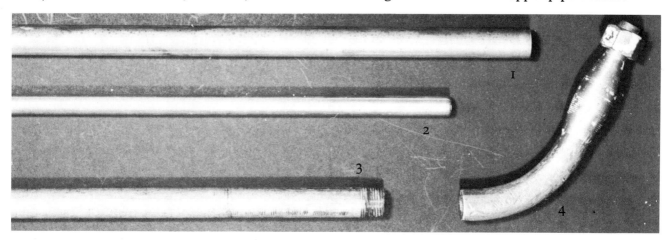

There are several types of metal pipe used in domestic plumbing, of which the following are most likely to be found: 1 22mm copper pipe, 2 15mm copper pipe, 3 ½in galvanised steel pipe with threaded ends, 4 ½in lead pipe. Steel and lead pipe will probably only be found in older properties as they are no longer in common use. You may also come across chromium-plated and stainless steel pipe.

stainless steel is a harder metal.

Do not confuse stainless steel pipe with chromium-plated copper pipe, which is even more expensive and can only be used on straight runs since you cannot bend it.

Other metal pipes

Lead was the traditional material for pipework and may still be found in older homes. It is not used today, however. To work it and make joins with it requires a high degree of skill and is not a job for the home handyman.

You may come across galvanised steel pipework in the home. Sometimes known as steel barrel pipe, it comes in two main sizes – $\frac{1}{2}$in and $\frac{3}{4}$in nominal bore. This is a confusing size system, since the outside diameter of the $\frac{1}{2}$in is about $\frac{7}{8}$in and that of the $\frac{3}{4}$in about $1\frac{1}{16}$in.

The screw threads are also the size of the outside diameter and on most of the galvanised pipe fittings they are tapered for a good joint.

You are not recommended to try and work with this type of piping since you will need a pipe vice and stocks and dies to cut the threads.

Bending pipe

Although it is slightly more difficult to bend pipe than to include angled fittings into the pipework at the appropriate places, it is much cheaper, usually quicker and allows a better flow of water through the pipe. And the flow of water can be important when the pipe is supplying a low pressure tap or a shower fitting.

You should be able to bend small amounts of 15mm copper pipe over your knee and, using more strength and annealing the copper (see below), you can also bend 22mm copper pipe in the same way. But to ensure accurate and consistent shaping – and when there are numbers of bends to be made – you will find it is worth your while to hire a bending machine.

You will need the correct size bending spring, depending on the internal diameter of the piping you are fitting. Remember here that Imperial size springs cannot be used in metric size pipe.

Before inserting the spring and bending the pipe (see pictures), make sure the spring is well greased with petroleum jelly. Not only will it be easier to get it in and out of the pipe, but it will also prevent the spring from rusting and getting stuck in the pipe. The easiest way to remove the spring is to overbend the pipe slightly then ease it back to the required angle. This will help release the pipe's grip on the spring.

Sometimes you may have problems getting the bending spring out of the pipe. You can often free it by inserting a small screwdriver into the loop at the end of the spring and twisting the spring clockwise as you pull on it.

As a last resort, try tapping all round the pipe where the spring is jammed with a wooden mallet to free it. This should not be necessary, however, if you bend the pipe correctly.

If you have any difficulty in bending the copper pipe, you may have to anneal – or soften – the copper first. If you have already inserted the spring, remove it by twisting it clockwise and pulling the loop at the end towards you. Heat the area to be bent with a blowtorch until it glows red and then immediately immerse it in cold water.

The biggest problem when bending pipe is to get the bend in the right position. If you only need one bend in a length of pipe, make sure you allow extra pipe at either end so that you can trim off any excess as required once you have checked the bend on the pipe run.

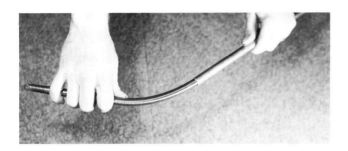

To bend pipe, first insert the greased spring so it is within the section to be bent.

Pull both ends towards you evenly and move the pipe to either side to increase the radius.

Slightly overbend then adjust the pipe before twisting the spring clockwise and pulling it out.

With a bending machine, insert the pipe with the former in place and pull on the handles.

If you have to make two bends in one length, try where possible to keep them a reasonable distance apart. If they are too close together, you will have trouble making the bends and removing the spring.

If you have to have two bends close together, use separate lengths and join the pipes between the two bends, which will be near the end of each length of pipe. Having made the first bend, work out where the next bend needs to start by measuring up where the pipe run will go and mark this position on the pipe with chalk or a piece of masking tape. Use this marker when making the second bend, ensuring that this is made from exactly that point on the pipe.

It is possible, though not really advisable, to straighten out a bent piece of pipe and use it again. If you have to do this, anneal the bent section of pipe as already described (see above).

If you have to bend a pipe a long way from either end, secure a length of strong cord to the bending spring and measure how far into the pipe the spring will have to go. Mark this length on the cord with a piece of masking tape so you know how far in to insert the spring. You may need a length of thin rod or dowel to push the spring along the pipe. When the bend has been made, pull on the cord to remove the spring.

Measure the length of pipe from the start to the finish of the first bend to be made. Mark on the pipe where the second bend will start, measuring from the end of the first bend.

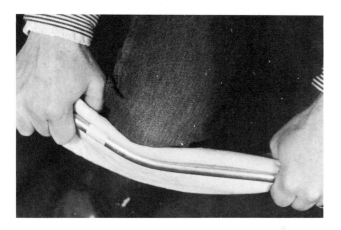

Having measured the length of the second bend, you can make this, working from the middle of this length. It is important to ensure you have enough pipe after the bend to pull on.

You may need to wrap the end of the cord round a small bar or screwdriver to get enough purchase to extract the spring.

Capillary fittings

Most capillary fittings are made of copper. Where brass is used, normal or dezincification type fittings are available (see pages 9–10). They work on the principle of drawing molten solder into the joint between the fitting and the pipe. This sets as it cools, giving a strong watertight joint.

This type of fitting is considerably cheaper than the equivalent compression joint and in most cases is easy to install. Once made, the joint is intended to be permanent and should not be disturbed. You can get union couplings incorporating a nut fixing for those joints that may have to be undone for maintenance purposes, for example on the WC cistern.

Inside the standard fittings there is a pipe stop, which ensures the right amount of pipe goes inside the joint. You can also get slip fittings. These do not have a pipe stop and are particularly useful when you want to break into the existing system or repair a burst section of pipe (see pages 41–43).

Here you simply cut out the damaged section of pipe, slide the slip fitting over one cut end, align the other cut end and slide the fitting back over that pipe as well before making the joint.

There are two basic types of capillary fitting – the Yorkshire and the end-feed. With the Yorkshire joint, the solder is already in the fitting in the form of a ring halfway down the joint (see picture).

The end-feed joint has no integral solder and you must add this as the joint is made. First clean the contact surfaces of the joint and the

These capillary fittings work on the principle of integral solder flowing round the joint as heat is applied to the join.

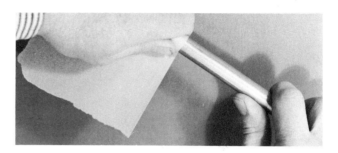

When making a capillary joint, first clean up the cut pipe end with glasspaper or steel wool.

Apply the flux to the outside of the pipe and the inside of the fitting.

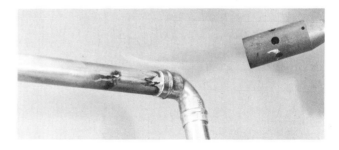

Heat both the compression fitting and the pipe itself until a complete ring of solder appears at the join. Allow the new compression joint to cool before cleaning it up with glasspaper or steel wool.

pipe with fine glasspaper and then apply the flux to both surfaces. Heat up the joint with a blowtorch, then remove the flame and apply the solder round the joint. If you keep the flame on, the solder will melt too quickly and drip.

Keep applying the solder until you have formed a ring round the mouth of the fitting. Repeat this procedure for each joint.

Sometimes you may only need to make one joint on a fitting, for example when you are planning to add to the joint after installation. In this case insert short pieces of untreated pipe into the other joints and wrap wet rags around them while you make the joint you need to prevent the rest of the solder melting as well.

One of the problems of making joints in situ is that the heat necessary to do the work could damage nearby surfaces, such as walls or floors, and create a fire risk. When soldering joints in this situation, you must use an insulating mat

When making an end-feed joint, apply flux to the cleaned end of pipe and inside the fitting.

Apply heat to the pipe and fitting until the flux starts to melt.

Remove the heat and apply the solder. If it does not flow, remove it and reheat the joint.

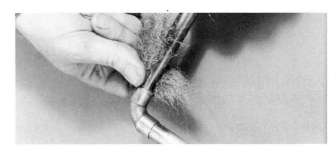

Apply the solder until a complete ring appears round the fitting. Clean the joint when cool.

behind the joint as protection against fire.

If you find a capillary joint starts to leak or you have not formed a complete ring of solder at the mouth of the fitting, you can add more solder to effect the necessary repair. But the area round the joint must be dry, clean and fluxed or the solder will not flow into the joint.

Compression fittings

This type is available as a manipulative or non-manipulative fitting. In the case of the former, the end of the pipe has to be specially shaped, so non-manipulative fittings are much easier to work with and more widely available.

You can buy the fittings in brass or in a dezincified material (see pages 9–10). These fittings have hexagonal nuts, whereas the normal brass ones have octagonal nuts.

Each fitting incorporates a socket with a pipe stop to ensure the correct amount of pipe fits into it. A compression ring (sometimes known as an olive), which has tapered ends, fits closely over the pipe and is then squeezed by the nut to ensure it grips the pipe firmly.

If you make this type of joint correctly, you should not need to use any jointing compound. However, if the joint starts to leak, you can cure

When making a compression joint, insert the pipe into the fitting up to the pipe stop.

Tighten the nut by hand until the compression ring grips the pipe and stops it rotating.

The range of compression fittings includes those non-manipulative ones made from a dezincification resistant copper alloy (top) and the more traditional brass ones (above).

Hold the fitting with one spanner and tighten the nut with another about two-thirds of a turn. If this join is not completely watertight, you can tighten it further with a spanner.

this by undoing the joint and wrapping PTFE jointing tape round it (see pages 36–38).

To guarantee a well-made joint, you must make sure that the pipe is completely round and not damaged or dented in any way, that it is inserted right up to the pipe stop in the fitting and that the component parts of the fitting are square to the pipe. This means the nut should rotate freely before gripping the compression ring. Never overtighten the nut, although with stainless steel pipe it will have to be tightened more than with copper pipe.

One point to remember is that compression rings and nuts are not always interchangeable between fittings from different manufacturers.

Special fittings

Many manufacturers are now catering for the DIY market by producing special fittings and selling pre-packed kits for specific plumbing jobs in the home. The range is extensive, but here are included some of the more useful items.

Tap connectors These fittings screw onto the threaded base of the tap and have compression-type fittings which will accept standard pipe.

Kopex pipes These are specially formed and corrugated along their length to enable you to bend them easily by hand to fit round corners and awkward obstructions.

Push-fit connectors These plastic fittings simply push over the ends of pipes, which must be cleaned first, to connect two lengths together.

Tap connector kit This kit makes connections to fittings such as a sink, bath or wash-basin a lot easier than the traditional method. It eliminates any problems that may be encountered when you are trying to cut and bend a rigid copper pipe to fit accurately into a tap connecting union. Often the supply pipe will be fixed in position and the job of connecting up a tap in situ can be very awkward.

The tap connector kit comes in three parts: a tap union which is screwed onto the bottom of the tap tail; a Kopex flexible copper pipe which is corrugated so it can be bent easily into the required position; and an Acorn push-fit connector to join the Kopex pipe to the supply.

When fitting a tap using one of these kits, first fix the tap in position using the necessary plastic washers and smearing non-setting mastic around the join. Secure it in place by tightening the back nut firmly. Wrap PTFE jointing tape clockwise around the bottom of the tap tail.

Make sure you clean up the ends of the Kopex pipe, removing any burr, and push one end into the tap connector as far as the pipe stop and the other end into the pipe connector. Loosely screw the union nut onto the bottom of the tap tail and bend the Kopex pipe into a suitable shape so that you can make the join between the pipe connector and the supply pipe.

Mark where the pipe stop in the connector comes on the supply pipe and cut the pipe there, removing any burr. Finally push the pipe connector onto the pipe as far as the pipe stop and tighten the tap connector union nut.

Tap conversion sets These give a modern look to an old-fashioned style tap. While retaining the base of the old tap, you replace the top and working parts. The taps are available with rising or non-rising spindles.

Tap connections With these kits, which fit onto an existing pipe, you can take a controlled

This connector kit with flexible pipe is to fit 15mm water supply pipe to a ½in tap.

Using the washer, fit the tap connector to the tap tail and tighten the union nut.

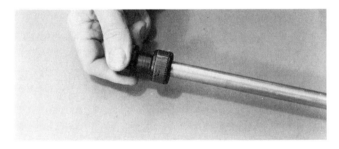

Having cut back the supply pipe, push the pipe connector firmly over the end of it.

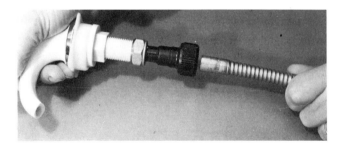

Bend the flexible pipe to the required shape and push it firmly into each connector.

feed off the existing system. They are particularly useful if you want to plumb in a washing machine or a dishwasher.

Some types are made to fit over a pre-drilled hole, while others actually make their own hole in the existing pipe as they are fitted. This saves you having to turn off and drain the system. Full instructions are normally supplied, but you may find the following points helpful.

First turn off the water supply and drain the pipe you are working on. Mark the position of the mounting bracket on the wall, drill and plug the fixing holes and screw the bracket in place. This will support the pipe while you mark the position of the hole to be drilled in the pipe by tapping lightly with a centre punch. This must be centrally placed.

Drill the hole of the recommended size, making sure you do not drill through the other side of the pipe. Measure how far in you have to drill and mark this distance on the drill bit with a piece of tape. Clean off any burr and fit the tap body, with the sealing washer in place.

Turn off the new tap and turn full on the tap at the end of the pipe you have drilled into. Restore the water supply, when any debris left in the pipe should be washed out through the open tap.

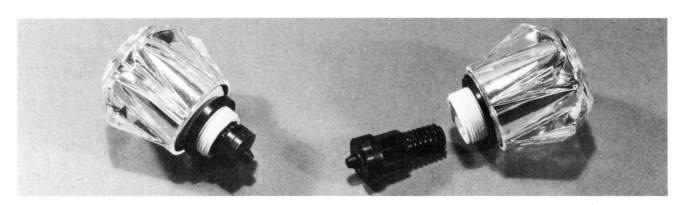

The advantage of a tap conversion kit is that you can change the design and working parts of an old tap without taking it out of the fitting.

You simply remove the headgear and screw in the new shrouded head. The kits are available with rising and non-rising spindles.

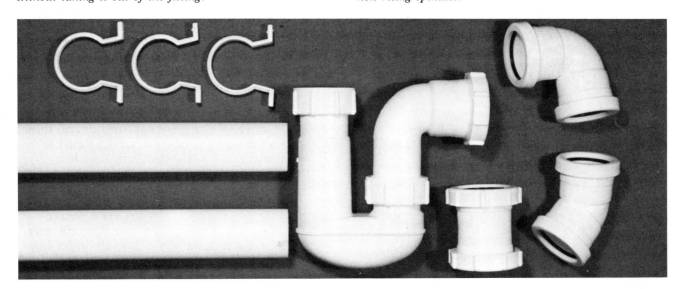

This is a typical plumb-in kit for a washing machine and includes lengths of 40mm (or 1½in) polypropylene tube, clips, a deep seal 'P' trap, elbows and couplers. You can, of course, buy the items separately as required.

Check that the fittings are suitable for plumbing in the machine, depending on the type you buy, and that you have all the fittings you need for the installation before you start work.

To fit a self-drilling tap, secure the back plate and screw the shoulder over the pipe.

Screw the shut-off tap firmly home to drill the pipe and then tighten the lock nut.

Taps and valves

There is now an enormous range of taps to suit any style bathroom or kitchen and even the old-fashioned brass taps, which became obsolete, are now available again to match a room with traditional decor.

Wash-basin and sink taps are normally 15mm (or ½in) size (the equivalent of the pipe feeding them) and have a long parallel threaded tail to enable them to be fitted to the wash-basin or sink. Bath taps are normally 22mm (or ¾in) size, in styles to match those of the wash-basin.

Most modern taps have a shrouded-head handle, often made of plastic, which hides the gland and simply pulls off when you need to carry out any work on the tap (see page 25).

The Supatap design gives a controlled flow of water through the 'handle' and this type is particularly convenient since you do not have to cut off the water supply in order to change the washer (see pages 26–27).

Many of the latest taps still incorporate traditional working parts, with a spindle that moves up and down as the tap is turned on and off and an adjustable gland. One type is available with a non-rising spindle and a non-rotating washer; this mechanism reduces wear in the tap. The gland has been replaced by a simple 'O' ring seal.

Probably the most advanced taps today are those with no washer. The water is controlled by two very hard ceramic discs, which do not wear. The tap only turns through 90 degrees from off to on.

Pillar taps are still available for use in kitchen sinks and, because of the height of the outlet, enable you to fill buckets from them.

Gate valves are another type of tap and these are fitted into the plumbing system to enable you to control the flow of water to specific outlets. They are usually fitted below the cold water storage cistern, by the hot water cylinder and on either side of the central heating pump.

The design allows an unrestricted flow of water when the valve is open, but they are only suitable on those parts of the system under low pressure (that is from the storage cistern onwards) and should be left either fully open or fully closed.

You can get mixer taps for sinks, wash-basins and baths, which have a single outlet for both hot and cold water. In some cases the outlet arm swivels to enable you to direct the

This T-Plus connector fitting enables you to take a feed off an existing pipe, but this type of fitting does not have a control valve fitted.

Clamp the fitting round the pipe, making sure the gasket is in exactly the right place. Check the branch points in the desired direction.

Connect up the branch installation, then check that the tap at the end of the installation is switched off to stop the water flow.

Remove the plastic cap on the fitting and tap sharply with a hammer. A loud pop indicates the connection to the existing pipe is made.

water in different directions – and this is useful if you have a double sink since the one water outlet can serve both sinks.

If you are considering fixing a mixer tap to the kitchen sink, where the cold water supply is direct from the mains, you must make sure that the mixer has a divided spout, so that the hot and cold water run separately to the nozzle via individual channels.

This prevents the high pressure cold water from the mains pushing the low pressure hot water from the storage cylinder back up its supply pipe. It also prevents the mains water supply becoming contaminated with household hot water should there be a drop in pressure or a mains failure.

Bath mixer taps are designed to fit in place of the standard bath tap outlets and these are single channel mixers since both hot and cold water are normally supplied at the same pressure – and not direct from the mains.

Combined with this mixer you can get a shower head attachment, which has a separate on/off control.

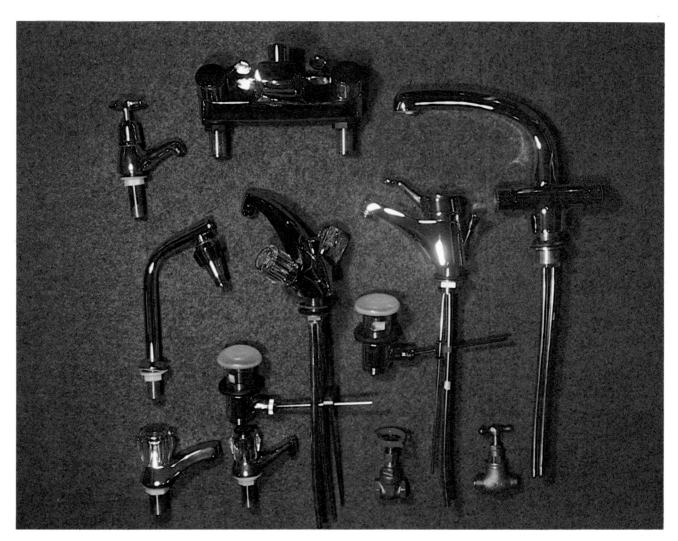

There is now a wide range of metal taps for use with sinks, baths, wash-basins and bidets and the choice will be determined to a large extent on the style and function you require.

You will have to decide whether you want single taps or whether you would like to fit a mixer tap. Taps, of course, include the brass gate valves and stopcocks (bottom).

5 Plastic components

In recent years the technology associated with plastic tubes and fittings has made tremendous advances. Forms of plastic have been used for cold water supplies and waste systems (see Chapter 6) for some time, but now certain types of plastic have been officially approved as suitable for hot water supplies as well.

Before you install this type of system, however, check with your local water authority to make sure the type you are considering using is acceptable.

You can now buy tubing and fittings in several different materials, but these are not compatible and should never be mixed. So make sure, if you do install this type of system, that all the components – tubes, fittings, adhesive, etc – are from the same manufacturer.

If you want to connect up sections of plastic tubing to the existing copper or steel system, you can get special connection fittings for this.

Most plastic tubing is flexible to a degree, which makes handling and installation that much easier. But even so, it should not be stressed or clipped to form bends – except in the case of some polybutylene tubing (see below). If in doubt, always check with your supplier.

Do not forget that plastic tubing will expand as it gets hot. With PVC tubing, for example, the rate of expansion can be as much as 6mm ($\frac{1}{4}$in) per 3m (7ft) length, so you must make an allowance for expansion in the system to avoid stress on the joints and other fittings.

Because plastic tubing is not rigid like copper or steel, you will need to fix it in position with special clips at maximum intervals of 500mm (or 20in) on horizontal runs and 1m (or 3ft) intervals on vertical runs.

Another point you should remember is that, unlike metal, plastic does not conduct electricity. So you cannot use this type of tubing as an earth return from any part of the wiring system. If you are replacing any part of the metal piping with plastic, make sure that if this has been used for earthing purposes an alternative earthing method is substituted to replace the existing one.

PVC tubes

Unplasticised PVC (uPVC) tubes are available only for cold water services in sizes from 11mm (or $\frac{3}{8}$in) to 22mm (or $\frac{3}{4}$in) nominal bore. The $\frac{1}{2}$in and $\frac{3}{4}$in sizes are compatible with 15mm and 22mm sizes. This type of tubing, which is grey in colour, is sold in straight lengths and cannot be bent. Solvent-weld fittings (see below) must always be used with this tubing.

You can also get a thinner-walled white PVC tubing in 22mm (or $\frac{3}{4}$in) size, which is used for overflow pipes only. It is not suitable for cold water services, because it is not thick enough.

Chlorinated PVC (CPVC) tubing is now available in 15, 22 and 28mm (or $\frac{1}{2}$, $\frac{3}{4}$ and 1in) nominal bore sizes. This type has been approved by the National Water Council for use in domestic hot and cold water systems. Again, this is supplied in straight lengths and cannot be bent – and solvent-weld type fittings have to be used.

Polybutylene tubes

This type of tubing, which is flexible, comes in 15, 22 and 28mm (or $\frac{1}{2}$, $\frac{3}{4}$ and 1in) sizes and can be bought in straight lengths or coiled. When you bend it, make sure the bends are supported by clipping them to the adjoining surface. You

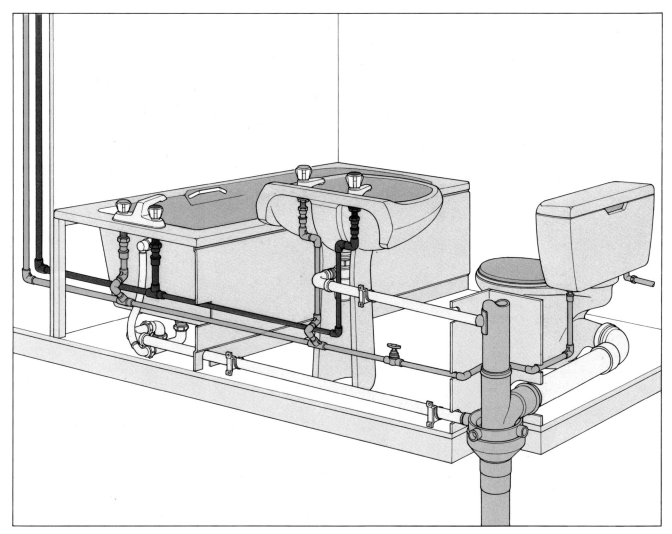

This is a typical bathroom layout using plastic tubing. The cold water supply from the storage cistern enters at the top left with the hot water on its right. Union nut connections are used at all the taps and at the WC cistern inlet. These type of connections enable you to carry out any maintenance, repair or improvement work required later without disturbing the rest of the pipework.

Plastic tubing has revolutionised plumbing installation. The range here includes (from left) overflow, three sizes of PVC, two sizes of polybutylene, and polyethylene tubes.

can use push-fit joints with this type, which can also be connected to standard brass compression fittings if you use a support sleeve inside the tube.

Polyethylene tubes

The first type of plastic tubing to be used in plumbing, it is available in 15, 22 and 28mm (or $\frac{1}{2}$, $\frac{3}{4}$ and 1in) sizes. Normally sold coiled, it is flexible and can be clipped into shape to form bends. When it is softened in boiling water it can be arranged in quite tight bends which will retain their shape when the tube cools.

There are special compression fittings for use with this type of tube. These are available in brass or dezincified material. You will need to use a copper support liner inside the end of the tube when tightening a fitting to ensure the tube does not collapse.

Polyethylene tube is black and cannot be painted. It is very resistant to frost damage and is normally used underground or for outside supplies where the tube run is out of sight.

Solvent-weld joint

As the name implies, this type of joint is made with a solvent which fuses together the tube and the fitting. Since solvents vary from manufacturer to manufacturer, you should always use the solvent recommended with a particular type of plastic tubing. Other types may not be compatible.

Since these solvents are very volatile, it is most important not to smoke near them and to ensure that there are no naked flames close by. Always make up the joints in a well-ventilated area and keep the solvent and cleaner out of reach of children. Make sure you replace the lid

When making a solvent weld joint, first cut the tube square. You can use a pipe clip as a guide.

Clean up the cut end of tube, both inside and out, with a half-round file.

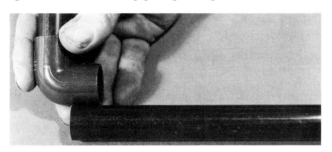

Check you have cut the correct length of tube by holding it against, not in, the fitting.

Clean the outside of the tube, roughen the surface and brush solvent on the fixing edge.

Following the same procedure, clean up the inside fixing surface of the fitting and carefully brush on the solvent. This is very toxic, so take care to avoid breathing in the fumes.

Push the tube into the fitting up to the tube stop, checking that the fitting is pointing in the right direction before the solvent sets. Hold the joint firm for about 30 seconds.

on the solvent and the cleaner immediately after use and never try to use the cleaner as a thinner for the solvent.

When making these joints, first cut the plastic tubing squarely with a hacksaw and remove any burr or rough edges with a file. Check you have cut it to the correct length by offering it up dry against the fitting. Never attempt to test the length of tube in the fitting. Since the fit is designed to be tight, you may have trouble getting the tube out again.

Make sure the fitting itself is not cracked or scratched internally and then with fine glasspaper slightly roughen the outside diameter of the tube and the inside diameter of the fitting. Using the recommended cleaning fluid, wipe the joint surfaces with a clean rag or paper tissue to ensure a good contact surface.

A word of warning here before you apply the solvent. This solution dries very quickly so you must make sure you know exactly which position the fitting should be in on the pipe before you try assembling it. You will not be able to alter it afterwards.

Coat the outside of the tube liberally with the solvent, brushing it on in a lengthways direction. Then coat the inside joint surfaces sparingly with the solvent. Never dip the end of the tube into the solvent. If the solvent gets inside the tube it will ruin the surface finish.

Immediately this is done, push the fitting over the end of the pipe with a slight twisting motion, align it in the correct position and then leave the joint for about five minutes.

You can flush cold water through the joint after an hour or so. You will need to leave the joint about four hours if it is a PVC fitting and the tube is carrying hot water. Wait at least eight hours before running water through at full pressure.

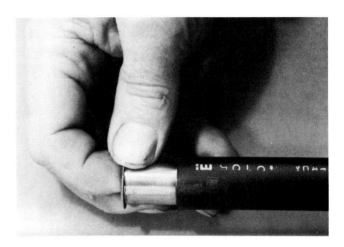

When making a mechanical joint, cut the end of the plastic tube square, clean off any burr and insert the liner into the end.

With the tube in position in the compression fitting, tighten up the nut by hand until the compression ring grips the tube. You should not use a spanner on the nut at this stage.

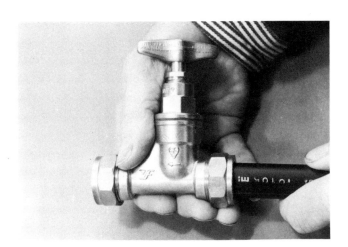

Loosen the compression fitting by unscrewing the nut and slide the tube into it as far as the tube stop inside the fitting.

To complete the joint, hold the fitting steady with one spanner and tighten the compression nut with another spanner. You should tighten this about two-thirds of a turn.

Mechanical joint

There are two ways of making up this type of joint on plastic tubing. You can either use the compression or push-fit method.

Compression This type of fitting is mainly used with polyethylene tubing, although it can be used with some polybutylene tubing as well. There are several different grades of tubing, so it is important that you check your compression joints are of the right size.

Cut the plastic tube to length with a hacksaw and file off any burr or ragged edges. Insert the supporting liner or sleeve (supplied with the joint) into the end of the tube right up to the flange on the liner. Then slide the nut and compression ring over the tubing and lubricate the outside of the tube with washing-up liquid or silicone lubricant.

Push the end of the tube right home into the fitting and tighten the compression nut by hand until the compression ring starts to grip the tube. Then tighten the nut a further one and a half turns with a spanner to ensure it grips the tube firmly. To check the fitting is tight enough, try twisting the tube; it should not rotate inside the fitting.

Push-fit The Acorn push-fit joint, designed for use with polybutylene tubing, is quick and easy to make. Unlike the compression fitting, this type can be rotated on the tube after the joint has been made, which can prove very useful in tight corners or awkward areas.

The seal in the joint is made with an 'O' ring, while a grab ring with teeth grips the tube tightly. These two components are separated by a washer. When necessary, you can change the components after the original joint is made.

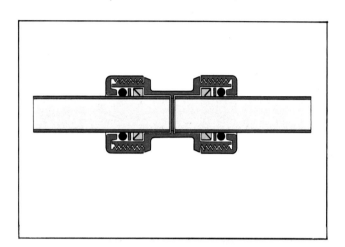

This illustration shows a section through an Acorn joint, with the metal ring that grips the tube and the rubber ring that makes the seal.

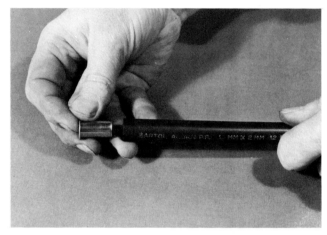

When making an Acorn joint, first cut and clean the end of the polybutylene tube before inserting the liner into the tube.

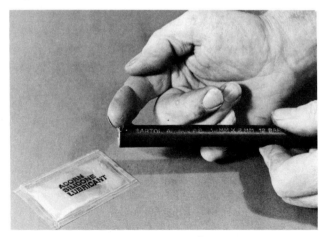

The next stage is to grease the end of the polybutylene tube. For this you will need a special silicone-type lubricant which you spread round the fixing edge of the tube.

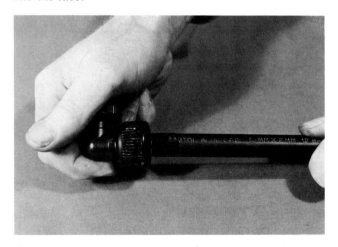

Finally push the tube right into the fitting where it will be gripped by a metal ring. The rubber 'O' ring makes the seal, while the tube can be rotated without fear of leakage.

Cut the tube to the correct length with a hacksaw or sharp knife and make sure the end is square and any burr or ragged edges are filed off. Insert the stainless steel support ring fully into the end of the tube. Then grease the end of the tube with a silicone-type lubricant and push the tube into the fitting as far as it will go to ensure a pressure-tight seal.

Plastic taps and fittings

With the present-day advances in technology, a full range of plastic taps for kitchen and bathroom use is now available. And they have received National Water Council approval for use with both hot and cold water.

They have several advantages over the traditional metal taps. For a start they remain cool to the touch even when controlling very hot water. They also incorporate a non-rising spindle and 'O' ring seal, which reduces the amount of maintenance required.

You fit them in place in the same way as the traditional metal tap, the only difference being that they are held in place with a plastic gland nut and not a metal one. One word of caution, however, when you are fitting this type of tap onto copper pipe. You should not overtighten the female brass coupling needed to join the plastic fitting to the metal pipe.

Wrap PTFE jointing tape round the plastic screw thread before making the join and hold the coupling still with one spanner while you tighten the compression nut gently with another spanner.

You can also get a range of special purpose taps in plastic. These have a variety of uses, such as controlling the water supply to a washing machine. Some of these have a spindle that only turns through 90 degrees from off to

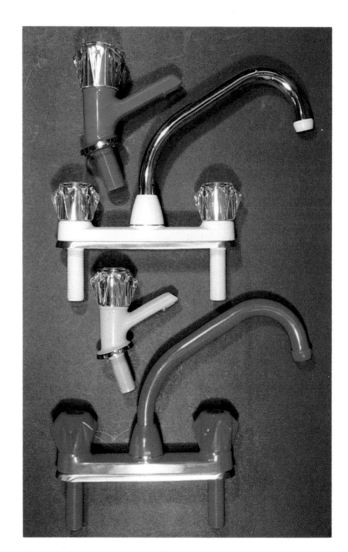

Plastic taps are now available in a range of styles for use with the different plumbing fittings and can replace existing metal taps since they are fitted in a similar way.

on. The range is similar to that of specialised metal taps (see pages 55–57).

You can get valve conversion sets made in plastic to fit existing metal taps (see page 52) and of course there is a wide range of plastic handles that can replace old metal ones.

Plastic pipe clips

These are now used on most installations since they not only look neat but are also simple to fit. One advantage they have over the traditional metal clips is that they allow for any small movement in the tubing due to expansion caused by a change in temperature.

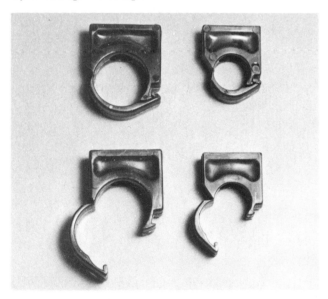

You simply screw these plastic pipe clips to the wall or solid surface, insert the pipe and lock it in. The clips shown here are designed to accept 15 and 22mm (or ½ and ¾in) pipe.

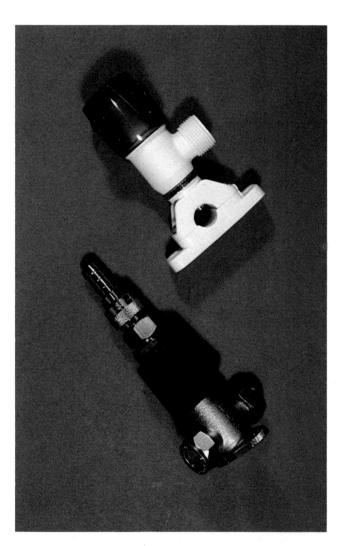

Plumbing-in taps are particularly useful when installing a washing machine, for example. The range includes the self-drilling plastic tap (top) and the Kontite thru' flow valve.

6 Waste fittings

Traditionally copper or lead piping was used to take the waste water from sinks, wash-basins, baths, etc. The cost of these materials would today be prohibitive when used for this purpose and there is now a full range of plastic tubes and fittings available. This includes the 22mm (or ¾in) nominal bore overflow pipe and the 32mm (or 1¼in) and 40mm (or 1½in) nominal bore waste pipes, up to the 82·4mm (or 3in) and 110mm (or 4in) nominal bore soil pipes.

These pipes are made from either modified unplasticised polyvinyl chloride (muPVC), which is coloured grey, or unplasticised polyvinyl chloride (uPVC), which is coloured white. Normally the white uPVC tubing is used only on overflow systems.

The fittings, in muPVC or uPVC, are of the solvent-weld type (see pages 58–62) and a wide range is available here as well. Included are adaptors to British Standard Pipe (BSP) threads for steel pipes, adaptors to copper pipes and, in the larger sizes, adaptors to cast-iron and pitch fibre pipes.

There is a further range of tubes available in polypropylene, which is coloured white, and the same sizes are made as for other plastic tubes.

Produced mainly for the DIY market, polypropylene tubes are slightly cheaper and do not require such a high degree of accuracy in installation. The fitting of this type of tube is made easier by the use of a joint ring inside, which provides the seal as the tube is installed. Polypropylene cannot be solvent-welded.

The advantages of the ring seal are that it allows for small errors in calculations of length and also for any expansion to be taken up in all the fittings.

If you are using this type of plastic, however, manufacturers recommend you should not mix it with PVC in the same installation.

Overflow pipe

This is the smallest waste pipe available and is fitted to take any overflow from cisterns and baths. As already mentioned, it is made of white uPVC, has a nominal bore of 22mm (or ¾in) and comes in 4m (13ft) lengths. The fittings available with this type of tube include the normal elbows, tees and sockets.

You can get tank connectors – either straight or with a 90° discharge – and adaptors to connect to 22mm (or ¾in) BSP fittings or copper pipe. Special clips are also available for this type of system.

Remember that uPVC tube is not suitable for carrying prolonged flows of hot water.

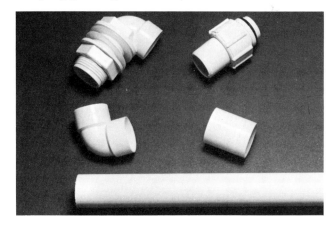

Overflow pipe is made from white uPVC, with a nominal bore of 22m (or ¾in). A range of fittings is available when connecting up this pipe, including a connector to the cistern.

Sink, wash-basin and bath waste pipe

You can use one of two types of plastic for these waste pipes – either muPVC or polypropylene (see above).

muPVC Baths and kitchen sinks normally discharge waste water through 40mm (or 1½in) nominal bore pipe and smaller wash-basins through 32mm (or 1¼in) nominal bore pipe. With the temperatures normally involved, PVC has been fully approved as a suitable material for this type of installation.

Make sure you use muPVC fittings with muPVC pipe. They are solvent-welded into position to ensure a watertight seal. The full range of fittings includes elbows, bends, tees, couplings, etc.

If you look at these fittings, you will see that the elbows and tees are made to angle of 91½ degrees. This ensures that all pipes that are nominally run horizontally are in fact installed at a slight angle – that is, 1½ degrees. No waste pipe should ever be fixed horizontally in case there is a gradual build-up of sediment deposit. The waste must always run freely at a slight angle down the pipe.

Where pipework may have to be disconnected at a later date – for example, if you get a blockage in the sink trap – you can get union couplings or ring seal joints to enable you to dismantle that part of the system.

You will find that if hot water runs through this type of waste pipe for any length of time – for example, when a bath or washing machine is discharging – the tube will expand and alter in length.

To give you some idea of the amount of expansion, when subjected to an increase in temperature of 39°C a 4m (13ft) length of

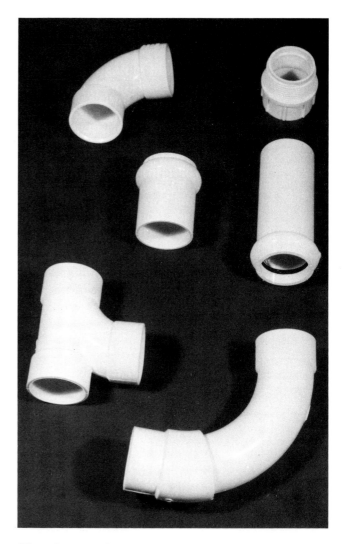

There is a complete range of PVC solvent-weld fittings to enable you to plumb in a waste system using plastic components. The choice of fittings will depend on the job in hand.

muPVC will lengthen up to 10mm (⅜in). Of course with short lengths this rate of expansion is not serious. With long runs, however, you must fit expansion joints every 1·8m (6ft) along the waste pipe.

You must support this type of tube with special PVC clips. On horizontal runs you are recommended to fit the clips at 500mm (20in) intervals. On vertical runs of pipe, the distance between clips should be about 1·2m (4ft).

You may sometimes have to run waste pipe round a bend that is not of the same angle as the standard fittings. If this is the case, you can get an adjustable bend fitting that will cope with angles varying from 90 degrees to 155 degrees. This special fitting is two-piece and you can cut it at different points along its length to give the correct angle and then solvent-weld the two sections together. And the fittings are available to connect with the appropriate BSP thread or copper pipe.

You can cut muPVC tube with a hacksaw. When you have made the cut squarely, clean off any burr with a half-round file.

Polypropylene You can buy this type of tube in the same sizes as muPVC. A little heavier than muPVC, it is available in white for all normal applications and in black where water is continually discharged at temperatures of up to 100°C. You will find it has a slightly greasy natural finish to the surface.

You can easily cut this type of tube with a hacksaw. You may get strands of polypropylene hanging from the cut ends; just remove these with a sharp knife.

As already mentioned (see above), you cannot solvent-weld this plastic. So the fittings

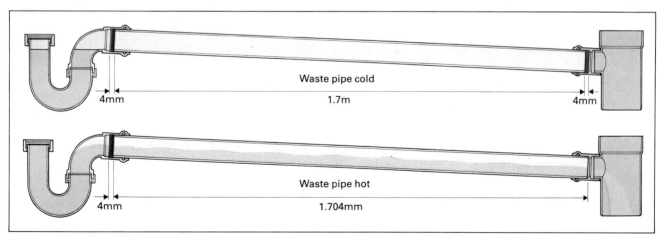

The maximum permitted length of 32mm (or 1¼in) waste pipe from an unvented trap on a wash-basin is 1.8m (or 6ft) from the trap to the stack pipe at a fall of 1½ degrees. The length must be shortened as you increase the fall. The gap at each end in the ring-seal joints is for expansion, which as you can see here occurs when hot water runs through the pipe and warms the plastic.

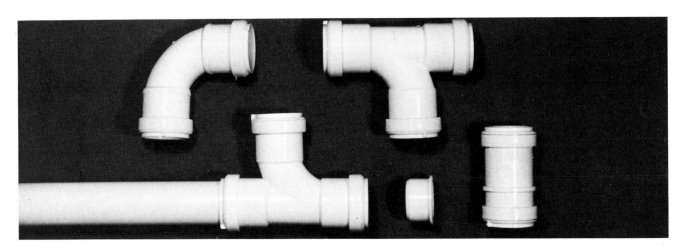

Polypropylene tube and joints cannot be solvent-welded and connections will either be of the ring-seal or compression type. Again a full range of fittings is available.

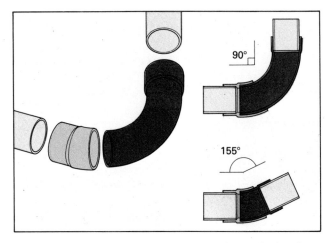

The advantage of the adjustable bend fitting is that it can be cut anywhere within the shaded area to give the correct angle when the end piece is solvent-welded into place.

must be of the ring seal or compression type. The fittings themselves come in a similar range to those made of muPVC, although you cannot get an adjustable bend in polypropylene.

One point to bear in mind is that if you are installing polypropylene pipe and fittings outside, you must protect them from the ultraviolet rays of direct sunlight. This can be done quite simply by painting them.

Soil pipe

This type of pipe, which takes water down to the underground drainage system, is of a larger diameter – normally 110mm (or 4in) nominal bore – and comes in uPVC. Generally it is coloured grey and is supplied in lengths of 3 or 4m (10 or 13ft). A range of fittings is available, including bends, branches, double branches, fittings with access and adjustable bends.

Because of the more solid nature of some of the waste in this system, all nominally horizontal runs must be at an angle of at least 2½ degrees and the fittings are all made accordingly to achieve this.

With this type of tube you can get either solvent-welded or ring seal joints, the latter being the most commonly used. The uPVC tube will expand according to the conditions and some allowance for this must be made (see page 58). This is usually achieved inside the ring seal joints.

You need to buy support clips to fit this tubing. With horizontal runs they should be spaced at 900mm (3ft) intervals and with vertical runs at 1·8m (6ft) intervals. You can also get special clips to support branch fittings on vertical runs of pipe.

You will find on modern waste fittings to the WC that there is a special socket and seal connection for the waste pipe. This is most convenient because it does away with the need for a cement seal. There are different types of fittings to suit all the standard WC outlets.

Connecting to the soil pipe While on the subject of soil pipes, there are certain principles you must follow (see pages 17–18) if you plan to make any connections to a main soil pipe. And you will need to check out any plans with the local water authority first.

When you are connecting a sink or washbasin to the soil pipe, there is a maximum permitted length of waste pipe from the trap to the soil pipe. If you are planning a longer run of waste pipe than this, you may be able to incorporate a vented trap arrangement. In this case, you must consult a qualified plumber or your local water authority. And remember that if you increase the angle of the gradient, you

Plastic fittings now make work on the stack pipe a lot easier. The range shown here includes both ring-seal fittings and a solvent-weld fitting (bottom) and a length of stack pipe (top).

The plastic WC connectors available for joining the back of the pan to the soil pipe are either 90 degrees (left) or straight.

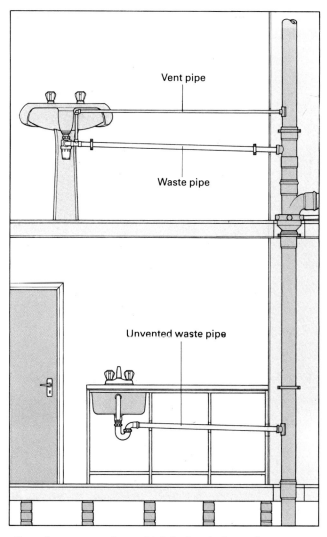

By using a vent pipe, which is fitted above the waste inlet to the stack pipe, you let in air to the waste pipe. This is essential to prevent the water trap from siphoning empty.

reduce the permitted length of the waste pipe.

You can make connections from the WC to the soil pipe using the recommended swept tee. Here, pipe runs must be kept as short as possible. There is a special collar boss fitting which allows you to connect the WC, bath and wash-basin waste pipes to the soil pipe at the same point.

When connecting pipes, of whatever size, to the soil pipe, there is a 'no connection' zone just below the junction. You must not make any other connections in this area, since otherwise the junction may get blocked.

Boss connectors are available if you want to make any new connections into an existing PVC soil pipe. These are solvent-welded into position.

This special collar boss enables you to join up the waste pipes from the WC, wash-basin and bath at the same point on the stack pipe.

This boss connector will be needed when you want to make a new connection to a PVC stack pipe. The boss connector has to be solvent-welded into position on the stack pipe.

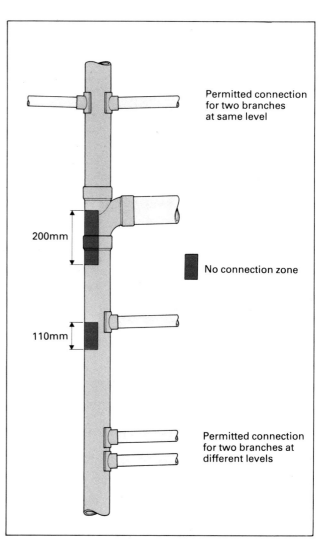

To avoid blockages or siphoning occurring in the waste system, there are restrictions on where connections can be made to the stack pipe. These 'no connection' zones are shown shaded.

Joining waste pipes

The method of fitting plastic waste pipe varies according to the type of tube you are using, as has already been discussed (see above). The three main types of fitting are the solvent-weld joint, the ring seal joint and the compression joint.

Solvent-weld joint First cut the pipe to length, using a fine-toothed hacksaw; make sure the cut is square. Clean off all external and internal burr with a half-round file. Before making the joint, you must check that the tube slots neatly into the fitting. Put the end of the pipe into the fitting and mark the alignment of the tube and fitting (see picture). Take out the tube and clean the inside of the socket and the outside of the tube with a clean cloth.

When making a solvent-weld joint, first mark the alignment of the fittings to ensure the pipes point in the right direction.

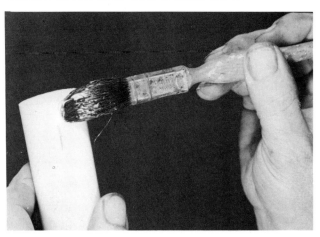

Clean up the end of the tube and then brush on the solvent carefully to the fixing surfaces only on the outside of the tube and the inside of the fitting. Do not dip the pipe or fitting into the solvent.

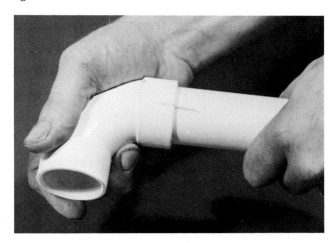

When you have brushed on the solvent, push the tube right home inside the fitting to complete the joint. Make sure as you do this that the pencil marks you have made align correctly.

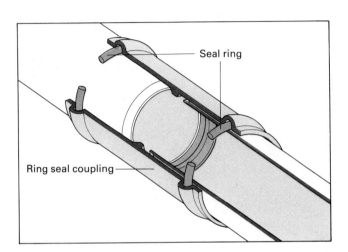

With a ring-seal joint, the tube slots inside the fitting up to the tube stop and a rubber ring effects an air and watertight seal.

When making the joint, cut the tube square and clean up the end before spreading a silicone lubricant round the outside fixing edge.

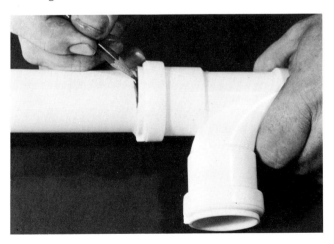

Having lubricated the end of the tube, push it right home inside the ring-seal fitting up to the tube stop. Then mark on the tube where it enters the fitting, using a pencil.

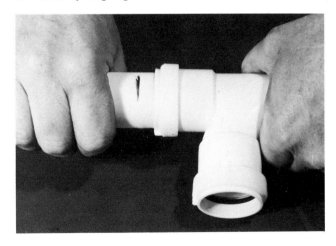

To complete the joint, pull the tube out of the fitting slightly so the pencil mark is about 10mm (or ½in) from the edge of the fitting. This allows for any expansion in the tube.

Apply an even coat of solvent over these two fixing surfaces. Never dip the end of the tube into the solvent, which can melt the inside surface. Push the tube right home into the socket, using a slight twisting movement, and check that the marks align.

Wipe off the surplus cement with a clean dry cloth and then hold the assembled joint in position for about 15 seconds to allow the solvent time to start setting. You can handle the tube and fitting assembly after a couple of minutes, but you should not use the system – that is, run water through the joint – for 24 hours.

Ring seal joint This is the easiest of the waste pipe joints to make and is probably the most practical since it allows for the pipe to expand. Inside each fitting you will see there is a tube stop to prevent the plastic tube going too far into the socket. Ideally you should allow a gap of 10mm (about ½in) or so between the end of the tube and the stop in the fitting to allow for possible expansion.

Cut one end of the tube square with a fine-toothed hacksaw and file off all external and internal burr with a half-round file.

Grease the end of the pipe with a silicone lubricant and push it right into the fitting up to the tube stop. Mark the end of the socket round the tube and then pull the tube out until the marked line is 10mm (or ½in) or so from the end of the socket. Make this allowance at each end of all lengths where they enter the fittings.

Compression joint You will find that compression joints on plastic fittings work on the same principle as those on copper fittings (see pages 50–52). But in this case the compression ring is made of rubber or plastic

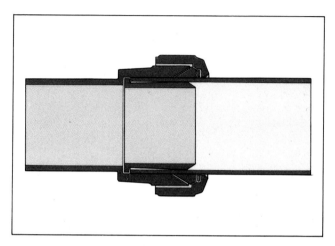

This plastic compression joint is used when you need to make a rigid but removable connection, for example to the stack pipe.

and the nut may also be of plastic. You must not use a spanner on the plastic nut since in every case this nut should only be hand-tight.

Bear in mind that these joints are not designed to allow for any movement in the waste pipe caused by expansion. Other means must be used every 1·8m (6ft) to cope with any expansion and this will require fitting a ring seal joint. As with all fittings, if you use a compression joint type, make sure that it is designed for the type and size of the tube being fitted into it.

Traps

The reason traps are fitted to all waste pipes is to provide a water seal between the room in which the particular fitting – be it a sink, wash-basin, bath or WC – is situated and the drains. Without this seal, the smell and bacteria present in all drains would penetrate into the house through the waste pipes – and, of course, health and hygiene would be at risk.

This is why traps must be incorporated to all waste systems, including the outside drains. And if you have the waste from your washing machine or dishwasher plumbed in, the waste pipe from these fittings must have a trap as well.

With the older two-stack drainage system, shallow waste traps were fitted and the depth of the water seal inside was no more than 42mm (or 1½in). With a single-stack system, where the risk of water siphoning is that much greater, these traps are not considered deep enough and here the depth of the water seal should be 75mm (or 3in).

You should fit this type of trap on all fittings in the house, since the cost is not great and the protection you get from the possibility of siphoning is much better.

The only exception may be on the bath, where because of space restrictions you have to fit the shallower trap. Because of the shape of a bath, with its comparatively flat bottom, the remains of water take a little time to run into the trap, by which time any water-siphoning caused by the pipe running full will have stopped.

With WCs and the outside drains, the traps are built in and, provided that they are correctly installed, generally you should not get any problems with them. The only trouble you might occasionally have is a blockage, due to foreign bodies getting trapped in the system or a build-up of sediment in the drains themselves. We have already discussed how to clear blocked pipes (see pages 43–44).

With sinks, wash-basins, bidets and baths, traps are not supplied ready fitted. You will have to buy these separately and install them into the waste system. There are two basic

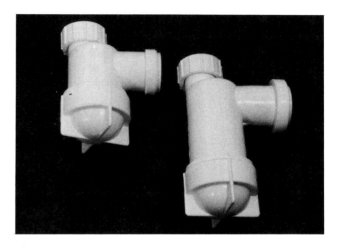

Here you can see the difference between a deep and shallow trap. The shallower type should only be used on the bath waste.

shapes of trap. With the 'P' trap the outlet is nominally horizontal and with the 'S' trap it is vertical.

The old-fashioned copper and lead traps are now virtually obsolete, except in some old houses, and have been replaced by a wide range of polypropylene traps. These are much easier to install and, most importantly, a lot simpler to clean in the event of a blockage.

One thing you must make sure of when installing a trap is that you use either a ring seal or compression joint when connecting it to the waste pipe so that you can remove it easily if necessary.

As their name implies, bottle traps resemble a bottle in shape and are available in 'P' and 'S' versions. There is a third version, known as a self-resealing or anti-siphon trap, which you should fit where you have a long or vertical waste pipe run.

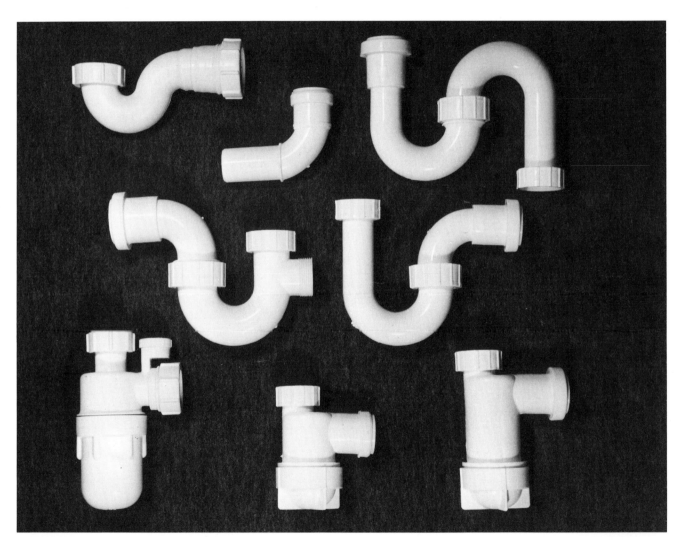

Plastic traps, which are now replacing the traditional metal types, come in a range of styles and sizes, depending on where they are to be fitted on the plumbing system. Of those bath, wash-basin and sink traps shown here, the top four are tubular and include a trap bend (centre) to convert from a 'P' to an 'S' trap. The bottom three are bottle traps, one (left) with an anti-siphon device.

This type is specially designed to overcome the problem caused when a large amount of water is released, causing the waste pipe to run full. With a normal trap, the water would be siphoned out of the trap and you would lose your 'seal'.

Bottle traps are suitable for use with the smaller fittings such as wash-basins. But they are not generally recommended for large fittings such as sinks, where the tubular-type trap copes much better with the increased volume and flow of water.

A seal depth of approximately 75mm (or 3in) should be adequate for all normal household requirements. The depth is measured from the top of the water to the highest point at the base of the seal (see diagram).

With these traps, there is a cap at the bottom which you only need to unscrew in order to clean the trap and clear any blockage (see pages 43–44). Make sure, of course, that you put a bucket or bowl underneath to catch the water in the trap.

Tubular traps are shaped like a 'U' bend and are available in 'P' and 'S' versions. As already mentioned, because of their design they can carry a greater flow of water than the bottle-type. To get at the trap to clean it, you simply remove the first half of the bend.

Because, with a bath, space limitations can prove a problem, you can get traps with either a 75mm (or 3in) or 42mm (or 1½in) seal depth. Some traps are designed to take the overflow as well. This is a lot safer if a tap is inadvertently left on when the plug is in, since the simple type of slotted waste outlet would not handle a large volume of overflowing water.

If there is a floor immediately below the trap, you should fit the type that incorporates a horizontal cleaning eye to ensure you can get reasonable access in the event of a blockage.

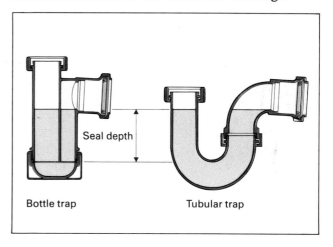

This diagram shows you how the seal depth is measured in both bottle and tubular traps. It is calculated from the top of the water seal to the highest point at the base of the seal.

7 In the kitchen

With the advances in modern technology and its associated labour-saving devices, it is as well to give some thought to the complete kitchen layout when the replacement of any major item becomes necessary. A double sink, for example, can make life much easier when preparing vegetables and washing up. Equally automatic washing machines and dishwashers will take much of the drudgery out of routine household chores.

Here we look at the techniques involved in replacing existing fittings and plumbing in some new ones. For virtually all the jobs there is no need to obtain any local authority approval, except if you are planning to plumb in a waste pipe – for example, from a washing machine or dishwasher – to the main stack pipe. In this case, check out the work you want to do first.

Fitting a mixer tap

The cold water supply to the kitchen sink is taken direct from the mains at high pressure, while the hot water supply will come from the storage cylinder at low pressure. You must therefore make sure that the mixer you get is specifically designed for the kitchen sink and not of the bath or wash-basin type.

The sink mixer has two separate channels in the nozzle of the tap to prevent the high pressure cold water forcing the low pressure hot water back up its supply pipe. Also, in the event of a mains pressure failure, it will prevent the possibility of the hot water contaminating the main water supply, where otherwise the hot water could force its way down through the cold water supply pipe.

The sink mixer tap comes in two basic designs – one that fits into the two existing tap holes in the sink and another that fits into a larger diameter central hole.

Whichever type of mixer you use, you will first have to drain off the water from the hot and cold supply pipes. In the case of the cold water, simply turn off the main stopcock just inside the house and open the draincock above it – making sure you put a bucket or large bowl underneath first. This will release the water between the main stopcock, the kitchen sink cold water tap

Remember when you buy a mixer for the sink that it is designed for use in the sink, since it must have separate channels for hot and cold water. The mixers have either one or two tails and your choice may be determined by the design of the sink. With one-tail mixers, check what size pipe they take – either 10 or 15mm – since you may need a reducing fitting in order to fit them to existing pipework.

and the cold water storage cistern in the loft.

To empty the hot water supply pipe, turn off the gate valve at the bottom of the cold water storage cistern that feeds the hot water system and drain off the water below this by using the hot water draincock, if one is fitted. This is normally below the lowest hot tap in the system. Again remember to place a bucket or large bowl underneath to catch the water.

If there is not a draincock on this system, turn on the kitchen sink hot water tap and a hot water tap above it, such as in the bathroom.

Should you not have a gate valve fitted by the cold water storage cistern, you will have to switch off the boiler or immersion heater and drain the entire hot and cold water systems from the lowest taps on each system. In this case, there will be no hot water or cold water to operate the WC flush until you have fitted the tap and refilled the systems.

Twin-hole tap Having drained off the water, unscrew the two union nuts below the sink that connect the cold and hot water pipes to the taps. Next unscrew the two tap unions from the bottom of the tap tails. Loosen the unions first and then unscrew the two back nuts above them. This will enable you to lift the taps and remove the unions.

While you are removing the nuts you will have to hold the taps to prevent them rotating. You will find a crowsfoot spanner very useful when you do this. Take off the back nuts and lift the taps out of the sink. Then clean out the holes with a screwdriver and glasspaper to remove any mastic or other debris.

Take the back nuts off the new mixer tap and put the fitting in position in the existing holes. Provided that the tails on the mixer tap are the same length as those on the old taps, you will not have to alter the pipework at all.

Some taps have a plastic or rubber gasket supplied, which is fitted between the tap and the sink. Smear a little non-setting mastic around the joint and the hole in the sink before putting the mixer tap in position.

If the sink is made of a thin metal, such as stainless steel, you will probably have to use a plastic 'top hat' washer between the sink and the back nuts to prevent these nuts becoming thread-bound before they are tight. Fit these washers with the square end against the sink and screw on the back nuts loosely. Do not tighten them at this stage.

Wrap a short length of PTFE jointing tape around the bottom end of the tap tails and fit the tap connecting unions. Tighten these fully with a spanner. Lift the tap tails over the ends of the two pipes and tighten the back nuts fully to hold the mixer tap firmly in the sink.

Check that the ends of the pipes fit comfortably into the tap tails. The compression ring already fitted should sit tightly against the union body. If it does, simply tighten the union nut by hand and finish off with a suitable size spanner.

If for any reason you cannot get a good fit between the tap tails and the existing pipework, you will have to cut off a section of the pipe and fit a new piece of exactly the right length. You will normally have to use a slip connector to enable you to fit the new length of pipe without disturbing the rest of the pipework.

When you measure the length of pipe, make sure you have allowed for one end of the new length to fit into the tap union right up to the pipe stop. You will also need a new compression ring for the old nut.

When you are satisfied the connections are sound, turn the water supply back on and allow

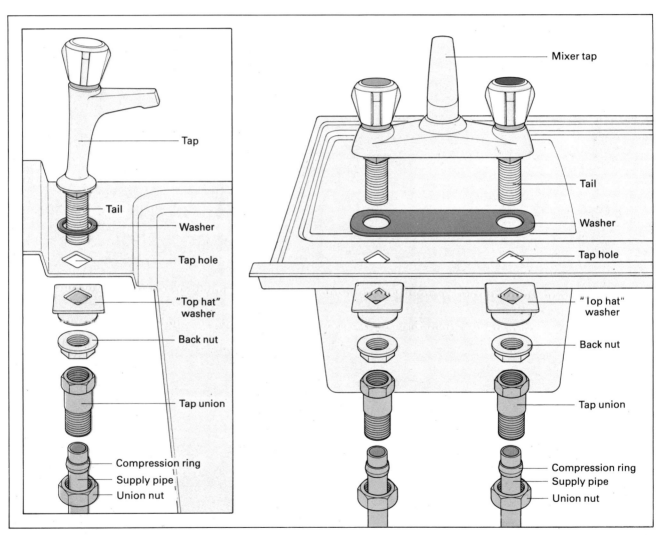

Here you can see the existing tap assembly, which you will need to strip down, having first turned off the water supply. When you have removed the old taps, you can install the new mixer tap into the sink. The exploded view of the fittings shows you the various components used in securing the mixer tap to the sink itself and making the pipe connections. Existing pipes can be used if the tap tails are the same length.

it to run freely through the tap until all the air bubbles have been removed. Then turn off the tap and check that neither of the tap unions is leaking. If you find you have a leak around the compression joint, you can try moving the nut another half-turn to stop the leak. If this does not work, you may be able to rectify the problem by remaking the joint using PTFE jointing tape or a jointing compound.

A better way would be to replace the old compression ring, if this was retained, and fit a new piece of pipe – as you would have had to do if you had failed to get a good fit initially. In either case, make sure you turn off and drain the water supply first before undoing the joint.

Single-hole tap This type of mixer has a single tail with two copper pipes of 15mm (or ½in) diameter protruding from the bottom. Only one hole is needed in the sink and this must be larger than the normal hole, which takes a 15mm (or ½in) tap.

If you are buying a new sink and fitting a new tap to this, suitably positioned holes should already be made to take either type of mixer. If not, you can enlarge one of the existing holes with a small half-round file – if the sink is made of stainless steel.

The other hole can be blocked up with a plastic blank cap. You could use this hole for a hot rinse brush attachment, which is available with some makes of mixer tap and carries installation instructions with it.

To fit this type of mixer, smear some non-setting mastic around the hole and the joint ring, insert the tap in place and position the 'top hat' washer and back nut. Tighten this nut fully and then gently separate the ends of the two feed pipes coming from the base of the tap.

You will, of course, have to fit new lengths of pipe to the existing supply pipes in order to make the right connection to the mixer tap. The method is the same as for the twin-hole mixer. Always use compression joints when making the connections so that you can disconnect the fitting later should any maintenance work be needed. And before making the connections, check that the two supply pipes are joined to the correct inlets – cold to cold and hot to hot (see page 10).

Finally turn on the mixer tap and restore the water supply, then allow it time to run freely until any air has been removed from the pipes. Check all the connections to make sure there are no leaks.

Bear in mind that there may well be some variation in fitting instructions for certain taps. Always check the manufacturer's instructions first.

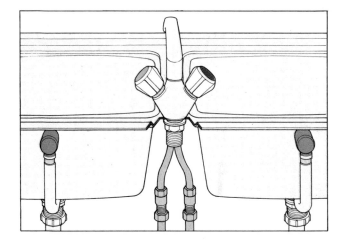

This single tail kitchen sink mixer tap is fitted to a double sink. The two pipes coming from the tail which are connected to the supply pipes may be of 10 or 15mm diameter, depending on the make.

83

Replacing a sink

Provided that you can use the existing waste pipe junction into the stack pipe, you will find it is a relatively straightforward job to replace an old sink with a new one.

Normally, the only time this may not be possible is if you are fitting a much deeper sink or if the new sink is lower than the existing one. In either case, you must check whether the waste pipe can still maintain a minimum $1\frac{1}{2}$ degree downward gradient from the trap to the stack pipe. If it does not, you will have to make a new junction in the stack pipe (see pages 105–107).

One point here you must not overlook when replacing a sink is whether there is an electrical earth connection to the metal piping. If there is, you must either leave it intact or replace it if you have to disturb it.

If you are fitting plastic tube in place of the existing metal pipe, consult an electrician as to where else the connection can be made since this earth has an important safety role.

Most modern sinks are supported by a surrounding cabinet or inserted into a hole in the working surface. In the case of the latter situation, it is advisable to let your supplier fit the sink into the work surface before delivery, since he will be able to guarantee a watertight seal around the edge of the sink.

Sinks are usually made of stainless steel nowadays – or of steel which is then vitreous-enamelled. Other types are available with plastic surfaces, but with these you should first check that the plastic surface will withstand the heat from pans straight from the cooker without suffering any damage.

Before you try to remove the old sink, you will have to disconnect the plumbing. Turn off the water supply as already described (see above). Then disconnect the tap unions and unscrew the union nut connecting the trap to the base of the sink. Check if there are any sink-fixing brackets and undo these before lifting the sink clear from the wall.

Put the new sink in position and work out what modifications, if any, will be needed to the existing hot and cold water pipes. It is much easier to carry out this work before fitting the sink, while there is plenty of room to work in.

Remove the back nuts from the taps and position each tap (or mixer tap) in the sink. Holding the tap union connector against the old pipe, work out the correct length for the connecting pipes, cut them to length and gently bend them out of the way.

An easier, but slightly more expensive method is to use a tap connector kit (see page

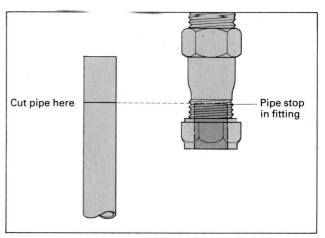

If you cannot connect direct to the existing supply pipes, you will have to cut them back and fit new lengths. Measure these to correspond with the tube stop in the connectors.

52), which has two tap connectors, two Kopex (flexible) pipes and two pipe connectors. You will find this kit very useful if access under the sink is a problem.

If you use this kit, cut the supply pipes considerably shorter. Because you are using flexible pipe, lining up the supply pipes with the taps is not necessary, although you should keep the flexible pipe run as simple as possible.

Having done as much of the preparation work as possible, you can now fit the sink into position using whatever method has been incorporated into the sink's design.

Make sure that the top edge of the sink is horizontal to ensure that the draining board has the correct slope on it. You can measure this with a spirit level. If it is not level, raise it as needed by using wedges under the unit.

The next stage is to fit the taps in place using the joint washers provided or some non-setting mastic. If you have a stainless steel sink, you will need to fit 'top hat' washers to prevent the back nuts from becoming thread-bound. Tighten these nuts firmly, ideally with a crowsfoot spanner, and wrap about 75mm (3in) of PTFE jointing tape around the bottom of the tap tails in an anti-clockwise direction. Fit the tap unions or connectors and tighten these firmly.

Slide a compression nut and ring over each supply pipe and spring it into position in the union. Each pipe must fit correctly when in position and the compression nuts should turn freely. Tighten them by hand and then give them a further two-thirds of a turn with a spanner; turn the nuts a bit more than this if stainless steel pipe is being used. Should you decide to use a tap connector kit, follow the manufacturer's instructions.

Having fitted the taps, you will need to

You can get combined waste and overflow kits for the sink. The type illustrated is designed without a banjo connection.

connect up the overflow outlet. This is normally fixed to the sink with screws and you should smear some non-setting mastic around the joint. You will need to do the same around the underside of the waste outlet before you position it in the bottom of the sink – over the rubber washer supplied.

To connect up the overflow to the waste, fit one of the large plastic washers, then the overflow banjo fitting and then another plastic washer onto the waste fitting. Smear a little non-setting mastic around the thread at the base of the fitting and then put on and tighten the large back nut. Wipe off any surplus mastic with a damp cloth.

To ensure the best possible conditions for the free flow of waste water, you should fit a tubular type deep seal trap to the sink. The most commonly used is a 'P' type trap.

If the existing waste pipe is made of copper

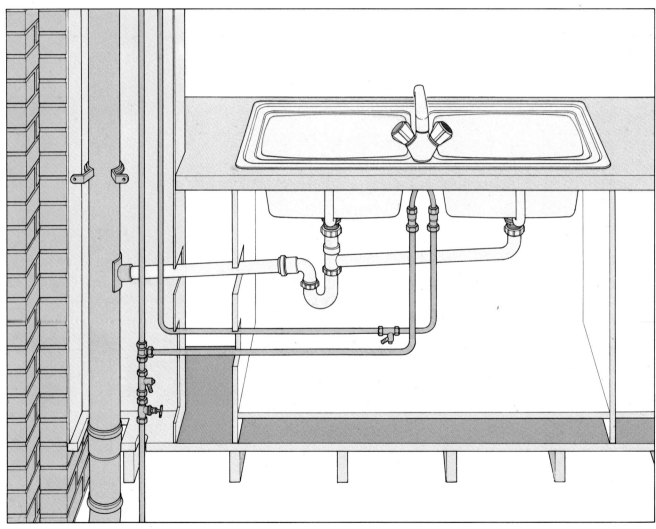

You will see from this diagram the connections needed when plumbing in a double sink. You can either fit a trap to each of the sink waste outlets or you can have both waste pipes incorporated into one trap, if you prefer. If you are using just one trap with a double sink, make sure this is fitted between the stack pipe and the nearest sink waste outlet to it to prevent sewer fumes feeding back into the kitchen.

or lead, check to see if it is fouled up inside, since this is a good opportunity to replace the whole waste pipe section with plastic fittings.

If you are fitting a double sink unit, you must make sure the trap is connected to the outlet nearest the stack pipe. In this case you then have to 'tee' in the waste pipe from the other sink into the waste from the first sink just above the trap. To make the plumbing-in for this situation easier, you can buy a double sink waste pipe set.

Fit the trap to the waste outlet, using hand pressure only to tighten the plastic nut. The waste pipe should fit into the ring seal outlet to within 10mm (or $\frac{1}{2}$in) of the tube stop in the trap to allow for possible expansion. To ease this pipe into place, use a silicone lubricant.

If the new sink is deeper than the old one, the trap will naturally be lower. This means that

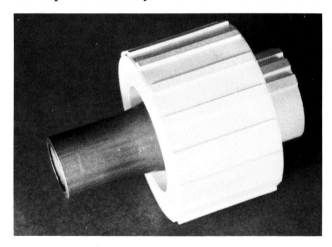

This special adaptor enables you to connect a PVC waste system to existing copper pipe. This will avoid the need to replace the complete system and is quite simple to fit.

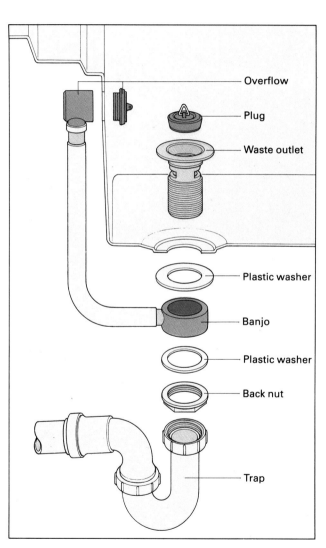

This exploded view shows the connections for fitting the waste and overflow pipes to the sink. The fittings required can be bought in kit form or you can buy them separately.

the waste pipe will have to be lowered to maintain the minimum downward gradient to the stack pipe. This may involve making a new hole in the wall or repositioning a bend in the pipe. Whatever adjustments are necessary, you must check that the waste pipe slopes at a minimum $1\frac{1}{2}$ degree gradient.

If you cannot make these adjustments, you will have to insert a new junction lower down the stack pipe and blank off the original one with a solvent-welded blank cap. Otherwise you will have to raise the height of the sink accordingly by making a plinth of the required height to coincide with the existing plumbing.

If the existing waste pipe is made of copper, you can buy a special adaptor to enable you to fit a plastic waste system. Bear in mind that you can probably sell the lead pipe to help pay for the new plastic components.

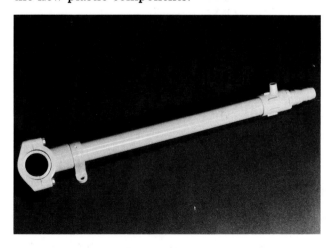

This plumbing-out kit provides an alternative waste system for a washing machine. It fits over the waste pipe and is useful when work is needed on the non-return valve.

Plumbing in a washing machine or dishwasher

The work required to install these machines is similar and the techniques involved can be used for either, depending on the basic requirements of your particular machine, details of which will be supplied with the equipment when you buy it.

Apart from the fact that all machines will need a supply of electricity, most will also need a hot water feed and normally you will want to connect them direct to the waste system.

As far as supplying electricity is concerned, you must have a 13amp socket suitably placed near the machine, which ideally should be reserved just for that machine. On no account should you operate the machine via a two-way adaptor while using another electrical appliance off the same socket. And if you have to install a new socket, keep it well away from the sink to avoid the possibility of water splashing up onto it.

To make life easier, most large DIY stores will supply a suitable pre-packaged kit of parts needed to plumb in a washing machine or dishwasher. But you must check first with the parts list to make sure the kit meets the requirements and situation of your particular machine.

If you are installing the machine in your kitchen, the final position may well be determined by the layout of the room. Obviously the most convenient place will be near the sink to keep the amount of extra plumbing to a minimum.

If you are creating a new utility area to house such a machine, it is a good idea to stand the machine in a sunken tray with a drainage hole feeding to the outside. In the event of a leak or a

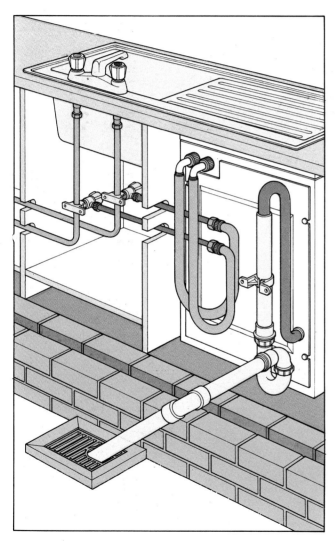

With this arrangement, the washing machine waste discharges into a gully. This is the simplest method, but a trap is still needed to prevent drain smells getting indoors.

hose breaking, which usually manages to happen at some time or other, the surplus water will drain harmlessly away without damaging floor coverings.

As far as connection to existing hot and cold water supply pipes is concerned, there is no problem about taking the hot water from the supply pipe to the kitchen sink. Most local authorities allow the cold feed to a washing machine or dishwasher to be taken from the mains supply, in which case you can plumb in to the supply to the sink tap. But you should check this out.

If you are not allowed to use this source of supply, you will have to plumb in the cold water from one of the feed pipes from the cold water storage cistern, and this may involve taking the supply from either a bathroom or WC pipe.

The easiest type of valve with which to make connections to the existing supply pipes is the self-drilling plumbing-in type that comes in kit form (see pages 52–53). You can use this valve for both the hot and cold water feed. Other valves are available which are fitted in a similar way to the existing piping, but they require more work.

Depending on where you are siting your machine in relation to existing pipework, you may find that the hoses supplied with the machine are not long enough to reach. In this case you will have to add a suitable length of copper pipe with a male-thread connector at the machine end and a female-thread union connector at the other end to fit the extra length to the valve. Make sure that this extension is firmly fixed with clips to a hard surface.

Before you connect up the hoses, make sure that the correct restrictor nozzles are fitted in the machine. Usually you should have a low pressure nozzle for the hot water feed and a high

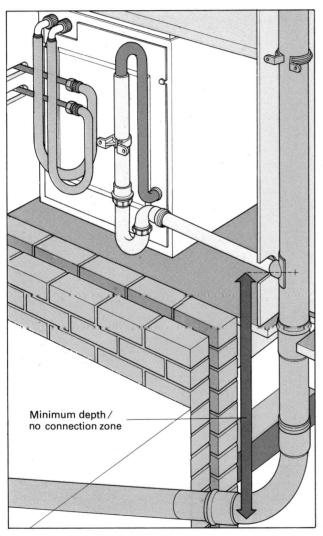

To avoid the possibility of siphoning back or blockages, there is a minimum permitted distance from the lowest connection to the stack pipe and the bottom of the drain – 750mm (2½ft).

pressure nozzle for the cold water feed, if this is being taken direct from the mains supply. Always check with the manufacturer's instructions.

If you look at the back of the machine you will see a waste hosepipe, which is usually hooked, situated at least 600mm (2ft) above floor level. When plumbing-in to take the water away through a waste pipe (rather than feeding the water into a sink), you must provide a break or air gap in the line to prevent the machine from siphoning itself empty.

The best way to do this is to fit an upright stand pipe of 32mm (or 1¼in) or 40mm (or 1½in) diameter to the wall behind the machine. The top of this pipe should be 600mm (2ft) above floor level. At the bottom of this pipe you must fit a tubular trap, which in turn should be connected to a waste pipe.

If the stand pipe is not long enough, you will find waste water is pushed back up the pipe when the machine is emptying and will spill over the floor. The waste pipe from the trap should then run, with a slight downward incline, to an outside gully. If there is no suitable gully, you will have to connect the waste pipe to the stack pipe (see pages 105–107).

Remember that you must get permission from the local water authority if you want to fit a new junction to the stack pipe. The lowest junction permitted is 750mm (or 2½ft) from the bottom of the drain (see diagram). This would normally be about 300mm (1ft) above ground level.

You may be able to use the existing sink waste pipe if this is at a suitable level and incorporates pipe of 42mm (or 1½in) diameter. To make a connection to this pipe, you will need a swept tee junction. In this case, make sure the sweep runs in the direction of the flow.

8 In the bathroom

The bathroom is one of the busiest areas in the home and it is therefore very important that it is designed to suit the needs of the people using it. Bathroom suites are relatively inexpensive and you may consider certain alterations are needed to ensure maximum efficiency.

Space is very often at a premium here, so the layout becomes crucial. Before you commit yourself to changes, therefore, you must work out carefully what the needs of the household are and how best to meet them.

If you decide to move any of the fittings, you should bear in mind what problems you may be creating when you come to alter the plumbing. In particular, take into account where the waste pipes are and whether it is practical to move them. You may feel you would like a shower unit added, but check first that the conditions in the bathroom are suitable for the type of shower you choose.

When deciding on making alterations in the bathroom, do not forget to allow for a certain amount of inconvenience, since at some stage in the work the water supply will have to be turned off. Particularly in the case of the WC, you must make alternative arrangements during the time it takes to complete the job.

Replacing a tap

Unless a tap has suffered excessive wear and tear, this is not a job you are likely to do unless you simply want a change of style in the bathroom. Plastic shrouded-head taps remain one of the most popular types, although depending on the decor or whether you are feeling in a particularly extravagant mood you can get brass or gold-plated taps. The choice of style is obviously yours – and there are plenty of designs from which to select.

The traditional tap mechanism, with adjustable gland and rubber washer, is still the most common, although there is a modified type with a non-rising spindle and 'O' ring gland and the most recent innovation uses ceramic discs in place of washers. With this mechanism manufacturers claim the tap has a much longer life since the seal improves with use – and there is the added advantage of just a 90 degree movement from the off to full-on position.

If the reason for changing a tap is simply a cosmetic one, you should consider a tap conversion kit, which replaces the working parts of the original tap with an up-dated mechanism and a new shrouded-head handle. This is a simple job to do.

If you are fitting a tap conversion kit, check the condition of the tap seat. If it is damaged or worn, you should first replace it using a nylon washer and seating kit.

Turn off the water supply to the tap (see page 25) and unscrew the old headgear nut. You may need to grip the body of the tap to prevent it rotating while you release this nut. If so, make sure you use plenty of padding between the tap and the adjustable spanner to avoid possible damage.

Having removed the headgear, check that the seat ring in the base of the tap is clean and not scratched or scored. If it is, you can get a new nylon washer and seating kit to overcome the problem. Then you simply fit the new tap headgear and handle and restore the water supply.

Replacing a tap on a bath involves basically the same technique – the difference with a bath tap is that it is larger and often less accessible and you will find the use of a crowsfoot spanner here will save you not only time but also skinned knuckles.

Before you start any work on the taps, turn off the water supply to them. You may find there is a gate valve on the feed pipes to the taps, but normally you will have to turn off the gate valves at the bottom of the cold water storage cistern in the loft (see page 25).

Turn on the taps to drain off as much water as possible from the pipes; you can also turn on other taps, for example on the ground floor, to speed this up. Remember, however, that the cold water sink tap will continue to run because it is fed direct from the mains and not from the storage cistern.

Unscrew the union nut connecting the feed pipe to the tap and loosen the tap connector union fitted to the tail. If possible, remove this union by springing the pipework clear of the tail. If you cannot do this, you will have to remove the union as you unscrew the back nut holding the tap in position.

You may have difficulty unscrewing this back nut over the old jointing compound used when the tap connecting union was originally fitted. In this case clean off the tail with a wire brush. Remove the old tap and check the length of the tails on the old and new tap. These should normally be the same. If they are not, you will have to alter the feed pipe accordingly (see pages 52-53 and 81).

Clean around the hole with an old screwdriver and wire wool, then position any sealing washer supplied with the tap over the tap tail. Smear non-setting mastic around the top of the tail and the hole. Put the tap in the hole and fit the back nut in position. With some thinner skinned baths, you will have to fit a 'top hat' washer above the back nut.

Wrap a length of PTFE jointing tape clockwise around the bottom of the tap tail and fit the tap connecting union. Tighten the union and back nut, making sure the tap is at the correct angle. Make sure the existing pipework will fit freely into the union right up to the compression ring, screw on the union nut by hand and tighten about two-thirds of a turn with a spanner to secure the joint.

Turn the new taps on and restore the water supply, then let the hot and cold water flow freely until any air has been cleared from the pipes. Turn the taps off and check for leaks.

Fitting a mixer tap/shower unit

As long as both the hot and cold water supplies to the bath originate from the cold water storage cistern, you can fit a mixer tap on the bath. You cannot use a mixer tap if the cold water to the bath is fed direct from the mains since you will run into the problem of having the two water supplies at different pressures – and the danger of contamination should there be a mains pressure failure.

The method of fitting this type of tap is similar to the standard tap. However you usually have to fit a rubber or plastic sealing washer between the tap and the bath.

You can buy mixer taps that include a shower attachment and these offer a very useful extra facility in the bathroom. From an economy point of view, the shower mixer will soon pay for itself by the reduced amount of hot water used in comparison to a bath.

In order to ensure the shower works efficiently under a reasonable pressure, you must not fit the shower head less than 1m (3ft) below the bottom of the cold water storage cistern. In some areas you may find there is a minimum size of storage cistern required if a shower is to be fitted. Check on this with your supplier before you buy.

The method of fitting the shower head will vary from model to model. Some, for example, are on a vertical rail and you can adjust the height by moving the head up or down as required. Others allow you two fixed positions, by means of brackets fitted to the wall, depending on whether a child or adult is using the shower. In each case the design eliminates the need to hold the shower head in your hand.

Before fitting any style shower, you must make sure that in its highest position the shower head is the required distance from the bottom of the storage cistern.

The shower head and mixer tap are normally connected by a flexible hose with screwed connections. Make sure that washers are fitted. To operate a shower mixer, first turn on the mixer tap and adjust the hot and cold taps until you get the desired temperature of water. Then turn on the shower control. The temperature

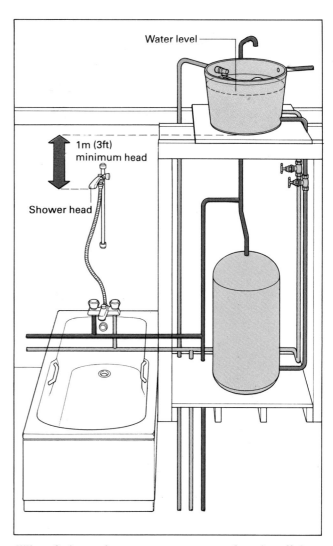

When fitting a shower you must ensure there is sufficient water pressure. This means that the shower head must not be less than 1m (3ft) below the bottom of the storage cistern.

may vary slightly when you do this, so check it again and make any adjustments needed.

There is one major problem with the shower mixer. When other low pressure taps are turned on in the house, this will affect the pressure and temperature of the water running through the shower head. This is particularly noticeable when the hot tap at the kitchen sink is turned on. If the shower mixer is on an upper floor, the shower water immediately goes cold.

Although this will come as a surprise to the person having a shower, it is not as dangerous as a sudden rise in temperature, which can occur when a low-pressure cold tap is turned on or a WC is flushed. The effect is to reduce the flow of cold water and therefore the shower water becomes that much hotter. It is therefore very important that the temperature of the hot water is kept low enough to prevent scalding and you should adjust the boiler thermostat accordingly. By doing this you will also reduce heating costs.

When you turn off the mixer tap controls to stop the shower – and you should always turn the hot tap off first – the shower control will automatically return to the 'bath' position. This avoids the possibility of getting a shower when you turn the taps on for a bath. It is, of course, much easier to switch the shower control to the 'bath' position rather than turn off the taps, which you can do when you have got out of the bath.

Replacing bathroom fittings

Baths and wash-basins have a long life if properly treated and it will only be an accident or a decision to change the style that will necessitate a replacement. If you decide to

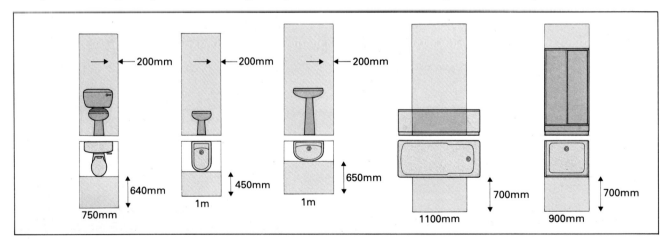

It is very important when redesigning the bathroom and moving the fittings that you make sure you leave enough space round each fitting to use it in comfort. This set of diagrams indicates the minimum space needed for each fitting and you should try to ensure your design keeps to this otherwise the bathroom will be inconvenient to use. Activity areas can overlap since you only use one fitting at a time.

change the style of your bathroom suite, then you should consider whether it would be an advantage to change the layout as well.

Modern bathrooms, particularly in smaller houses, tend to be very compact and nothing like the vast areas shown in the manufacturers' catalogues. But if you are changing all the bathroom fittings, then certain alterations to maximise the space available or the installation of an extra fitting, such as a bidet, are possible.

There are certain limitations as far as optimum space requirements are concerned (see diagram) and you should make sure that for comfort and safety these are observed whatever layout you choose for your fittings.

As far as changing round the bathroom is concerned, the biggest problem will be with the WC. The slightest movement of this fitting will involve alterations to the waste plumbing, which with the large diameter soil pipe is a major job. So this should be avoided.

You may, however, be able to reduce the amount of space taken up by the WC by replacing the existing one with a close-coupled model, which is far more compact (see picture).

The position of the waste pipes and their relation to the junction with the stack pipe will also, to a degree, affect the position of the bath and wash-basin. But the alteration to the plumbing is not a major problem, provided that you take into account the permitted length of waste pipe from the wash-basin trap to the junction with the stack pipe – that is 1·7m (5ft 8in) – and the minimum angle of downward slope – that is $1\frac{1}{2}$ degrees.

If you need to run pipe round corners, make sure this is done with sweep fittings to ensure minimum resistance to water flow. This is because all the waste water in a wash-basin or bidet will drain away immediately. If the waste pipe runs full, it will siphon out the water seal in the trap, letting sewer gasses into the bathroom.

If you need a longer pipe run than the maximum permitted, you can use a larger diameter pipe – 40mm (or $1\frac{1}{2}$in) – with a 32mm (or $1\frac{1}{4}$in) trap. This will enable you to increase the length of the run to 3m (or 10ft).

With a bath, the waste water runs away more slowly, allowing the trap to refill after the majority of the water has passed through it. The maximum permitted length of run for 40mm (or $1\frac{1}{2}$in) waste pipe from a bath trap to the stack pipe is 3m (or 10ft).

If you increase the length of the waste pipe from the bath, for example, you may have to raise the level of the bath to ensure you get the minimum fall on the waste pipe. You can use

The modern close-coupled WC, where the cistern sits at the back of the pan, is compact and therefore particularly practical in a small bathroom where space is limited.

this to advantage in certain cases by designing a 'split-level' bathroom, although you should bear in mind that the height from floor to ceiling should never be less than 2.27m (7ft 7in).

One way in which you can make more use of the space available is to replace the bath with a shower booth. If this sounds too drastic, you can get special tubs which take up far less space, fill an awkward area and can be used as a shower cubicle as well.

Baths are now available in a variety of materials. Although the original heavy cast-iron type is still available, the more common types now are of pressed steel or plastic. The final choice of material is one that you must make according to your needs or preference. Steel baths tend to be stronger, while plastic or glass fibre ones are warmer to the touch and available in a larger range of shapes and designs.

One point worth bearing in mind is that modern materials are less resistant to abrasives than the traditional vitreous enamel and you should be careful what type of bath cleaner you use in case you scratch or damage the surface.

When you come to replace any of the bathroom fittings, you will have to plan the work carefully to ensure that inconvenience to the household is kept to a minimum.

One of the major problems is that you will have to cut off the water supply to the bathroom. A useful way of overcoming part of this problem is to put the old taps back on the end of the supply pipes after you have removed the wash-basin or bath. Make sure, if you do this, that the taps are turned off and kept in the off position. You can then restore the water supply to the rest of the house so that other outlets, including the WC, can be used as normal.

It is a good idea to replace the WC first since this will probably retain its original position and therefore the plumbing-in work will be kept to the minimum. And, of course, the WC is indispensable and you will want to restore this facility as soon as possible. Check that you have all the correct fittings, materials and tools immediately to hand to save wasting any time and make suitable arangements with a neighbour in case of an emergency!

Replacing the WC

The WC is the most used and probably the most essential item of plumbing in the home. It should give long service and, cistern apart, should not create any problems. However, if it does get cracked or damaged, you should replace it and not attempt any repair work. You may, of course, want to change it as you replace other bathroom fittings to keep a matching suite.

The original type of WC with a high-level cistern is now obsolete, although you may come across this style in old houses. Various types of WC with low-level cisterns are available and in each case the cistern is fixed to the wall and has a short pipe connecting it to the WC. The most modern designs are known as close-coupled, where the cistern is supported on an extension of the pan and also fixed to the wall.

There is a choice of pan styles, too. The most common is the washdown, where the contents of the pan are washed down the soil pipe by the force of water fed out beneath the rim. The other, more expensive type is the siphonic. The principle of operation is that water from the cistern first causes a partial vacuum to develop in the 'U' bend, which sucks out the contents of the pan. The pan is flushed after it has emptied. The remaining water then refills the 'U' bend. The great advantage of this type of WC is that it is much quieter to operate.

While most WC pans are still ceramic, you can get plastic versions; the cisterns are also available in both types of material. Apart from their extra weight, ceramic models are more prone to condensation on the outside due to the hot, damp atmosphere in the bathroom, which temporarily gives them a matt appearance.

WCs used to be fixed to the floor with a cement mixture and were jointed into the ceramic soil pipe using a mix of cement and sand. When upstairs WCs were screwed to wooden floorboards, the movement of the boards caused the joint to crack. Putty was then used, but in time this also went hard and was prone to cracking and therefore leaking.

Nowadays the joint is made using plastic or rubber connectors which fit over the outside of the WC and down the inside of the soil pipe. This joint design obviously makes replacement that much easier to carry out.

When you buy a new WC, you must check that the one you choose will mate with the existing soil pipe end. There is a wide range of outlet designs, from 'P' traps with horizontal or angled outlets to 'S' traps with vertical outlets. The choice of WC connectors is also large and will, for example, enable you to fit a close-coupled WC in place of an older style low-level WC.

It is possible to cut a cast-iron soil pipe with a hacksaw, but you cannot alter the soil pipe if it is of salt-glazed earthenware. In this case you will have to adapt the position of the WC to fit the existing soil outlet.

When planning for a new WC, do not forget to take into account the distance from the back of the WC to the wall. With the standard wall-mounted cistern, there is a degree of flexibility and you must ensure that the pan is positioned

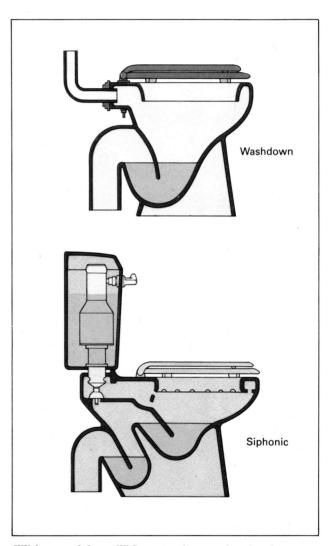

With a washdown WC, water from under the rim flushes away the contents. With the siphonic type, the contents are sucked out of the pan by a partial vacuum before the trap refills.

in such a way that the seat stays up when lifted. With the close-coupled type there is, of course, no adjustment. If, because of the existing soil pipe arrangement, you cannot site the new WC right up against the wall, you may have to use blocks of wood as spacers between the wall and the cistern or build out the wall by lining it.

Do not attempt to remove the old WC until you have got everything ready and to hand to fit the new one, since an open soil pipe is not the most hygienic thing to have in a bathroom. Time, therefore, is of the essence. In any case, you should block the soil pipe opening with a ball of newspaper until you are ready to connect up the waste.

The first job, as always, is to turn off the water supply and disconnect the old cistern at the valve union. Flush the cistern and soak up the remaining water inside with cloths. Remove the overflow and WC connecting pipes and take out the cistern.

To remove the pan itself, you must disconnect the joint with the soil pipe. If this has been made with cement or putty, try chipping it out carefully with a small chisel or old screwdriver. If this proves impossible, you will have to break the joint with a cold chisel. Make sure you wear a thick pair of gloves and protective goggles to prevent injury from flying pieces. While you are doing this, you must try to ensure that no pieces of ceramic fall down the soil pipe and that you do not break the pipe entry.

Unscrew the pan from the floor or break it away from the cement base with a bolster chisel, depending on how it was originally fixed. In the latter case, you will have to clean up the rest of the cement from the floor before you install the new WC.

As already mentioned, you should block up the soil pipe entry with a ball of newspaper and clean up round the edge. When you are ready to make the waste connection, you can remove the paper and clear out any debris. Fit the WC connectors, lubricating the joints with washing-up liquid, and then position the WC. Check that the rim of the pan is level; you can do this with a spirit level. If it is not, insert wooden wedges in the appropriate places under the pan.

You should fix the pan to the floor with brass screws, which do not rust. When you do this, it is a good idea to use plastic screw head washers to prevent any possible damage to the ceramic as you tighten the screws. With a concrete floor you will have to drill and plug fixing holes first. It is not normal practice now to grout round the base of the pan, although you can do this if you want to.

Before you fix the cistern in position, fit the

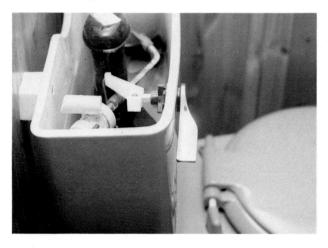

You may find, because of the position of the existing soil pipe, that there is a gap between the cistern and the wall. In this case use spacer blocks to help secure the cistern.

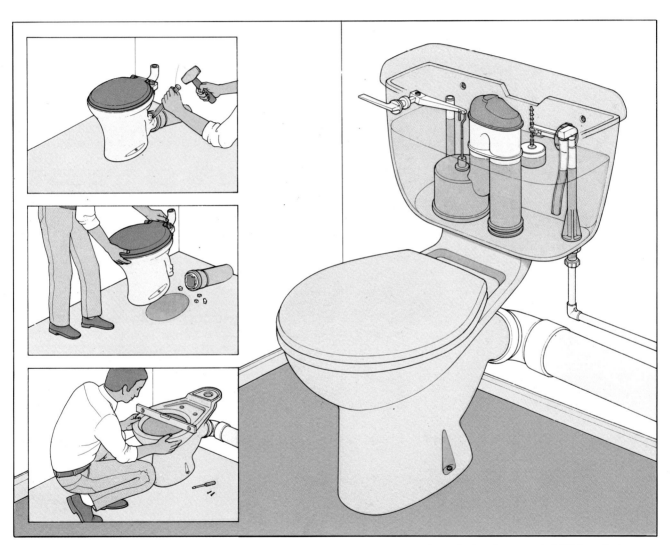

When replacing an old WC, you will first have to remove the existing pan and cistern. Here you can see the stages involved in breaking the connections to release the pan.

The section through the close-coupled WC and cistern shows where the connections have to be made and illustrates the position of the siphon tube and a Torbeck cistern valve.

siphon inside, using the washers provided to ensure you get a good, watertight joint. Tighten the back nut firmly to prevent any leaks. When you put in the siphon, check that the ball arm on the valve can move freely up and down when it is fitted later on. If not, move the siphon until the arm operates freely. Put on the flush lever and connect up the operating mechanism.

You can now fix the cistern to the wall. With the wall-mounted type, position it at the recommended height, drill and plug the fixing holes in the wall and secure it in place with stout brass screws. A good idea here is to use a similar size steel screw first to 'tap' the thread and then replace it with the brass screw. This will avoid the possibility of the softer brass screw breaking off halfway in.

Cut the ready-shaped WC connecting pipe to length. Fit the rubber seal over the back of the WC at the flush outlet, slide the pipe into place and tighten up the union nut. Then connect up the overflow pipe, making sure it runs at a slight downward incline through the outside wall.

If the level of the cistern has altered from the original position, you will have to block up the old hole in the wall using a suitable plaster filler and make a new one. This will have to be of a 25mm (1in) diameter to take the overflow pipe and you will need a long masonry drill for this, which you can hire quite cheaply. Alternatively, use a small cold chisel and a club hammer to make the hole. Having inserted the pipe you will have to fill the gap around it with plaster filler.

With a close-coupled WC, the cistern is fitted over the back of the WC pan with a rubber gasket, which will be provided. Tighten the two wing nuts below the cistern to hold it firmly in place. Fix the cistern against the wall using brass screws as before, but do not strain the cistern into place. If there is a gap between the cistern and the wall, fill it with blocks of wood and fix the cistern into these – or, as already mentioned, line the wall to bring it out to the back of the cistern.

Fit the overflow pipe into place on the opposite side to the water supply and run it at a slight downward incline through the outside wall, as before.

Next fit the ball valve in the cistern, but make sure you check that the correct nozzle is fitted – a small one for high pressure, if the cold water is mains-fed, and a large one for low pressure, if the supply is from the storage cistern.

Follow carefully the manufacturer's fitting instructions supplied with the valve you buy and make sure, with a close-coupled cistern, you fit joint washers at the base of the plastic tower. Put the Garston valve (see pages 31–32) in position and check that the float arm moves up and down freely. If it does not, you can move the valve by loosening the nuts on the bottom of the tower. Adjust the nuts on the steadying screw so that it touches the side of the cistern firmly.

Connect up the existing water supply to the bottom of the tower. Here it is best to use a plastic fitting on the plastic base of the tower.

Turn on the water supply and adjust the valve arm (see pages 31–32) until you achieve the correct water level in the cistern, when the valve will shut off the supply. Flush the cistern and check for any leaks. Check also that the correct plug is fitted to give either single or double flushing as required (see pages 34-35).

Replacing a bath

When replacing a bath, turn off the water supply to both the hot and cold taps (see page 36) and drain off the relevant part of the system. Take out the side panel and keep it carefully if you intend using it again. Remove the wooden support battens.

The next job is to disconnect the taps. Using a crowsfoot spanner, unscrew the union nuts connecting the pipework to the taps and, if possible, spring the pipes out of the tap connecting unions so that you can take these off the tap tails. Disconnect the waste pipe at the trap and, if the overflow runs out through the wall, cut through the pipe. You can remove this pipe when you have taken out the old bath.

Having disconnected the plumbing to the bath, the next problem is how to get the old bath out – particularly if it is made of cast iron, which is very heavy, and the bathroom is upstairs. The resale value of this type of bath is very little and the easiest solution is to break the bath into manageable pieces.

You can do this quite easily with a club hammer, but make sure you are wearing a strong pair of gloves and protective goggles to prevent the possibility of injury from flying pieces of bath. The best place to start is in the middle of one of the sides since this is the weakest point. When you have broken the bath into suitably sized pieces, you can carry these out of the house. The new bath should be supplied with feet and the necessary fixing brackets, so you can discard the old fixings as well.

Incidentally, new baths and wash-basins are supplied with thick protective sticky paper round the vulnerable areas to prevent them being damaged in transit or while being fitted. Do not remove this paper until you have installed the fitting.

If any major work on the supply or waste pipes is necessary, should you, for example, want to change the existing position of the bath, make sure you do as much of this as possible before putting the new bath in place.

You may be fortunate enough to find that little or no alteration to the pipework is required. But should adjustment be needed to the supply pipes, and particularly if these are fixed rigidly in position, the easiest way to extend these pipes is by using flexible tap connectors. In this case, you will have to cut off a suitable length of pipe and make these connections (see pages 81–83) before you position the bath.

When you come to put the bath in position, check carefully with the manufacturer's instructions with regard to the assembly of the sub-frame, for example, which will be supplied with plastic models. All baths will 'settle' after they have been filled with water and taken the weight of a person as well. So you must stand the feet or sub-frame on wide battens at least 25mm (1in) thick – and preferably hardwood.

Position the bath on these battens up against the wall and adjust the feet as required until the rim of the bath is at the desired height and level all round. You can check this with a spirit level on each side and end in turn. Make sure all the feet are taking the weight of the bath and tighten the lock nuts on them.

With the bath fixed in position, you can now put in the taps. If you have a pressed steel bath, you will probably need 'top hat' washers to prevent the back nuts from becoming threadbound. With plastic and glass fibre models, you will often find a stiffener is provided to strengthen the tap fixing and this should be positioned on the underside of the bath. Use the plastic or rubber washers provided between the taps and the bath. A smear of non-setting mastic on these will help provide a watertight seal. Then fit and tighten the back nuts.

Using plenty of padding to avoid damage to the surface, grip the tap with an adjustable spanner, wind a strip of PTFE jointing tape around the bottom of the tap tail in a clockwise direction and fit and tighten the tap connecting unions.

Spring the existing supply pipes into position under the tap tails. Check that each pipe enters the union as far as the pipe stop and sits comfortably inside. It must never be wedged in at an angle. Finally tighten the union nuts by hand and give them a further two-thirds turn with a spanner. Alternatively fit the flexible tap connectors in place. Do not turn the water supply back on just yet.

You can get a packaged waste kit, which includes the overflow fitting. Fit this first, smearing a little non-setting mastic around the hole in the bath to ensure a watertight seal. If the fitting is of the screw-in type, you can tighten it in place by putting two screwdrivers into the holes and rotating the fitting using another screwdriver threaded between them. Push the flexible hose onto the back of the fitting and onto the overflow pipe on the bath trap.

Smear a little non-setting mastic round the waste hole in the bath and place the outlet fitting in position. You should put a washer between the bath and the back nut before tightening this nut with a spanner. Then connect up the waste trap, which may be a deep seal type or a special bath trap with a 40mm (or 1½in) seal depth if space is limited. To ensure a leakproof seal, wrap a strip of PTFE jointing tape clockwise around the bottom of the waste outlet before securing the trap. The plastic nut should be tightened by hand enough to secure the fitting. If you overtighten it you may damage the plastic.

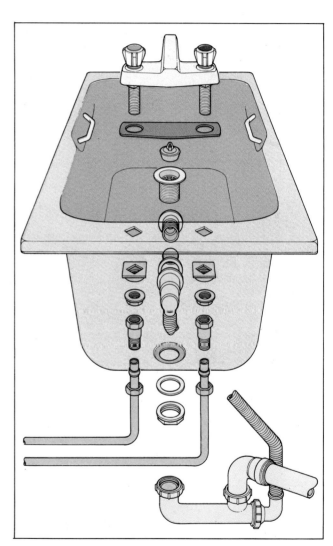

This is a typical plumbing-in arrangement for a pressed steel bath with a mixer tap. The overflow is now usually connected to the trap rather than fitted through the outside wall.

Do not forget that the waste pipe must run to the junction with the stack pipe at a downward incline of at least 1½ degrees. And if you are using plastic pipe, remember to allow a 10mm (or ½in) gap in the ring seal fittings to allow for possible expansion.

You can now turn on the water and check fo leaks. When you are satisfied that everything is working properly, you can start tidying up. This means stripping off the protective paper from the new bath and replacing the bath panel that fills in the exposed side. Fit this panel back carefully to avoid damaging it. The side support battens should be screwed to the wall and the top support batten then fixed to these, since you cannot make any fixings to the bath itself.

You must also fill in the gap between the edges of the bath and the wall. This is best done with a silicone rubber sealant. It comes in easy-to-apply tube form and in a full range of colours to match the colour of your bathroom suite or decor. This type of sealant forms a positive leakproof bond between the bath and surrounding tiles or wall and is flexible enough to withstand any expansion or contraction of the bath, as well as a limited amount of settlement.

Replacing a wash-basin

Wash-basins are available in three basic types: pedestal, wall-hung and countertop, which can be part of a vanity unit. With the pedestal and countertop types, the plumbing is concealed behind the fitting.

Although the wall-hung type allows greater flexibility of fitting, you will have to organise some way of concealing the pipework. You may decide to box it off, although this will spoil the lines of the room. Alternatively you can build a false wall section, for example using pine panelling, and run the pipes behind that.

Hanging brackets and their fixings vary, but full instructions on how they should be fitted will be supplied with the new wash-basin. The normal height for a wash-basin is 800mm (or 32in) from the rim to the floor. Of course you can vary this as you wish to suit the requirements of the household.

Traditionally wash-basins are available in ceramic materials, although plastic, pressed steel and stainless steel models are also available. Sizes and shapes vary enormously, from a small design to suit a separate WC where space is at a premium and the wash-basin is to be used for hand-washing only to those big enough to bath a baby in. You will have to choose according to your needs.

With ceramic wash-basins, the overflow is usually built in to connect with the waste. With

One way of hiding the pipework to a wall-hung wash-basin is to fit a small cupboard underneath. This not only gives you storage space, but also access to the plumbing.

some of the other types, however, separate waste and overflow kits are necessary.

The taps are normally of 15mm (or ½in) diameter size and conventionally the cold tap is situated on the right. You may need 'top hat' washers when fitting steel and plastic wash-basins, but these are not normally required with ceramic models, which are much thicker.

The waste outlet is 32mm (or 1¼in) in diameter and with some designs the waste pipe may have to run through a fixing bracket.

Bottle traps with a deep seal are recommended for use with wash-basins and these have a neat appearance.

The techniques involved in fitting taps, waste outlets and traps to a wash-basin are similar to those involved for the bath (see pages 100–103) and you can refer to these when tackling this job.

If you are installing a wash-basin into the top of a vanity unit or countertop, make sure you use a good mastic sealant to give you a watertight seal around the edge of the wash-basin. With a pedestal or wall-hung wash-basin, use a silicone rubber sealant to provide a waterproof seal at the back edge against the wall or wall tiles.

When you check for leaks on the supply pipes to a bath or wash-basin, make sure that the hot pipes and their joints are watertight when hot. And once you are satisfied that they have been plumbed in correctly, it is a good idea to lag them to save heat. Lag the cold pipes as well, since this will prevent condensation from forming on them due to the hot, damp atmosphere in the bathroom. Excess condensation can, in extreme situations, lead to rot around the skirting or floorboards.

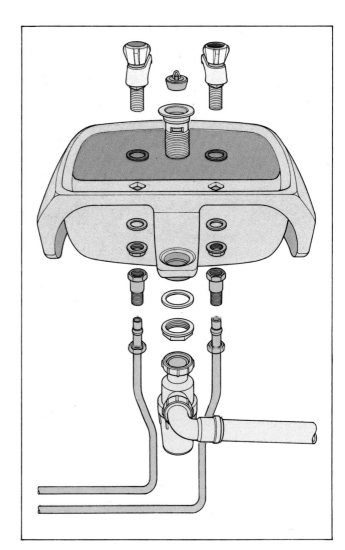

This is a typical supply and waste arrangement for a ceramic bathroom wash-basin. This type has the overflow duct built-in to the system. The normal waste size is 32mm (1¼in).

9 Adding to the system

Before installing any new plumbing and adding to the system, you should check with your local water authority to make sure the work you want to do complies with the regulations. These will vary from one area to another. For example, in some areas you will need a licence for an outside tap, while in others you will have to have a meter fitted and be charged accordingly.

If you want to install a shower, you may be required to fit a larger cold water storage cistern and in all areas you must get permission to make additional connections to the stack pipe.

If you undertake any of the following installations, you must first check them out with the local water authority.

Connecting to the stack pipe

You can get solvent-welded or strap-on connection bosses to make junctions in 110mm (or 4in) diameter PVC soil pipe; these bosses are of 35 or 40mm (or 1¼ or 1½in) diameter.

If a collar boss is already fitted that connects the WC, bath and wash-basin waste into the stack pipe, there may well be an unused junction boss on this fitting. In this case you can drill out the hole and insert the appropriate solvent-welded fitting in it.

Whichever arrangement you use, you will have a ring seal or solvent-welded connection to which you can fit the new waste pipe.

The first job is to check on the route the waste pipe will take and this must be kept as straight and short as possible. The maximum permitted length for 35mm (or 1¼in) diameter waste pipe from a wash-basin or bidet is 1·7m (6ft 8in) and the minimum downward gradient should be 1½ degrees. The wash-basin or sink should have a deep seal 75mm (or 3in) trap fitted. You can increase the length of the pipe run to a maximum of 3m (or 10ft) if you are using 40mm (or 1½in) diameter pipe, while still using a 35mm (or 1¼in) trap.

The reason for these restrictions is to prevent the waste pipe running full and siphoning the water from the trap, thus removing the protective barrier between the drains and the inside of the house. If you increase the gradient, you must reduce the length of pipe run. For example, with a gradient of 5 degrees, the maximum permitted length is reduced to 700mm (28in).

An additional precaution worth taking if your installation is near these limits is to fit an anti-siphon trap. This admits air during siphonic conditions.

This precaution is advisable because, due to the sloping sides of a wash-basin or bidet, the waste water in these fittings drains away immediately. If the waste pipe runs full, the water in the trap will be emptied and none will be left in the fitting to top the trap up again and restore the seal. With sinks and baths, which have a flat bottom, the remains of waste water will run into the trap after the bulk of water has passed through the trap – thus maintaining the water seal. With these fittings, the normal maximum permitted length of 40mm (or 1½in) waste pipe run is 3m (10ft).

These conditions also apply to washing machines with 40mm (or 1½in) diameter waste pipes.

If you find your waste pipe runs have to be longer than the maximum allowed, you can increase the length if you use a vented trap. This involves running a separate vent pipe from the waste pipe at the trap to a higher point on

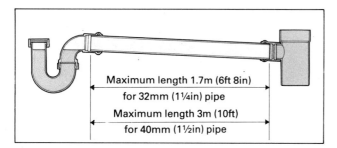

You must adhere to the maximum permitted length waste pipe to avoid the trap being siphoned.

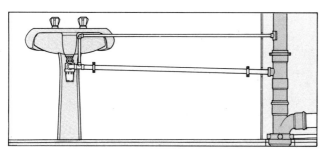

A vent pipe lets in air and prevents the trap siphoning empty when the waste pipe runs full.

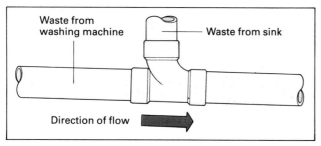

If you need to incorporate a tee fitting into a run of waste pipe, make sure that you use a swept tee fitting and that the sweep points in the direction of the stack pipe.

the stack pipe. This lets air into the waste pipe just after the trap, thus preventing the contents of the trap being siphoned out.

In this situation you should seek the advice of a qualified plumber or consult the local water authority.

When you come to install the waste pipe, you must make sure that any bends are swept bends and not sharp corners. And if you need to join two runs of waste pipe – for example, from a washing machine and sink – you must make the connection using a swept tee in the direction of flow.

When making any connection to the stack pipe, remember there are certain areas around a junction that, because of the risk of blockages or obstructions, are regarded as 'no connection' zones. Check on where these zones are (see page 73), but do not forget that if you connect up a new waste pipe close above and opposite an existing junction, you may inadvertently be creating a 'no connection' zone where this original junction exists.

When it comes to fitting a new junction to the stack pipe, you must follow the manufacturer's instructions since the method involved will vary depending on the make of components you buy.

A word of warning before you start the work. You must make sure that no-one uses a WC or other water fitting above where you are working.

You can make the necessary hole in the stack pipe quite simply either by using the correct size holesaw attachment to an electric drill or by drilling a series of holes in a circle slightly smaller than the opening required. First, however, drill two holes in the centre of this circle and loop a piece of wire through them. Hold onto the wire while you work to prevent the disc falling into the stack pipe as you cut it out. If necessary, cut round the holes with a padsaw to release the disc. Then carefully clean round the cut edge with a half-round file.

Having made the hole, you are then ready to fit the connection, following the manufacturer's instructions carefully.

Fitting an outside tap

This is a useful addition to the domestic plumbing system, particularly for watering the garden, washing the car or cleaning up those muddy boots before you come in the house. You will need permission to install one and it will almost certainly affect your water rates. Normally you can connect this tap from the rising main just after the main stopcock and draincock.

You can either get an outside tap kit or buy the components separately. Make sure the tap you fit has a screwed outlet to take a hose coupling.

At this stage it is worth considering fitting a stopcock on the feed pipe to the tap, inside the house. Not only will this enable you to control the supply to the tap, which may be necessary if the tap suffers from frost damage, but you can also turn off the supply to prevent unauthorised use of water outside.

The simplest way to fit this control is by means of a self-drilling tap, like the type used for washing machine and dishwasher connections, installed at the junction to the rising main (see page 89). From this tap you will have to feed the supply pipe through the wall.

The tap should either incorporate or be fitted to a back plate, which you must screw to the wall, first drilling and plugging the fixing holes in the correct place.

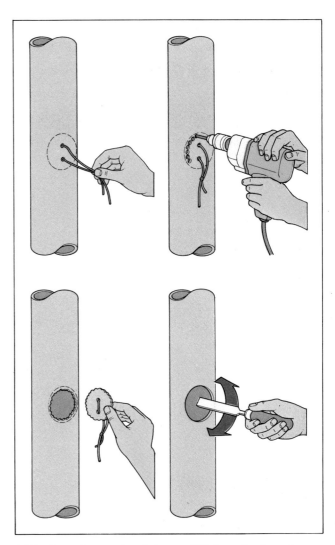

This shows how to break into a PVC stack pipe when making a new waste connection. It is important to ensure that the disc you cut out does not fall into the stack pipe.

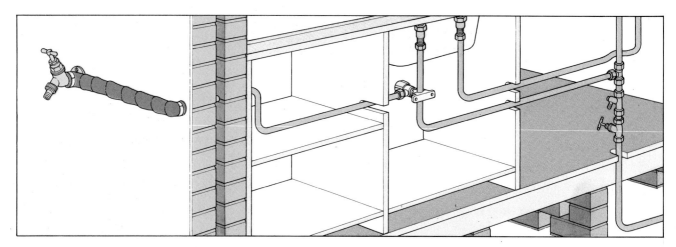

This outside tap installation incorporates a self-drilling tap as an indoor stopcock, with lagged copper pipe supplying the feed to the outside tap. You can use polybutylene tube instead to reduce the risk of frost damage which can occur in exposed situations. Make sure the outside tap is firmly fixed to the wall.

It is a good idea to use plastic pipe for the water feed, since this is more resistant to frost damage and therefore less likely to freeze up. Make the necessary connection to the taps at either end of the pipe run (see pages 60-66) and fix it to the wall at 500mm (20in) intervals using plastic clips.

Make good any damage to the wall round the hole with a suitable exterior plaster filler and then lag the outside of the pipe from the tap to the wall. You can do this with proprietary lagging or by boxing in the pipe (see pages 38-39) and filling this box with vermiculite chips (the sort used for loft insulation).

Installing a shower unit

There are several advantages of a shower which make this addition eminently practical. A shower is obviously ideal when space is at a premium and you are prepared to dispense with a bath. Equally you can combine a shower with a corner bath or tub.

Apart from being a more hygienic way of washing, since you are not lying in dirty water, it is quicker than running a bath and makes less demands on the hot water supply – and therefore costs much less.

If there is room to fit a shower booth as well as a bath, it means that two members of the family can wash themselves at the same time – an important consideration during that early morning rush or when you are getting ready to go out in the evening.

Apart from the bath mixer tap, which we have already discussed, there are three types of shower installation. You can fit a shower over the bath using the existing hot and cold water supply. Alternatively you can fit an instant shower; this is fed by just cold water which is

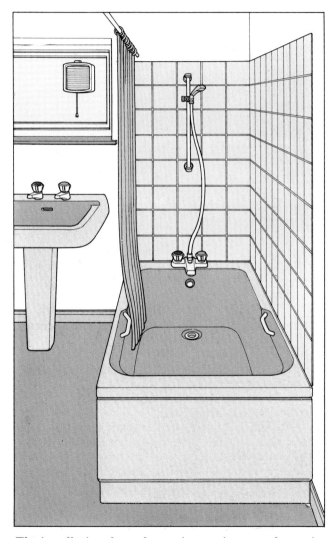

The installation shown here using a mixer tap shower is the simplest to fit in the bathroom. You must use it with care and check on the temperature of the hot water (see text).

heated as the shower is used. Or you can install a separate shower booth, either in the bathroom or, if space is limited, elsewhere in the house.

As already mentioned, you will need permission from your local water authority for all these installations and you may have to increase the size of your cold water storage cistern.

The amount of water you get through the shower head will depend to some extent on the model you buy – and does vary from manufacturer to manufacturer. If the rate of flow is less than a gallon (4·5 litres) per minute, you will only get a fine spray which will probably not be enough to keep the body warm in a cold bathroom. Most showers have an output of between one and one and a half gallons (4·5–7 litres) per minute, although you can get showers which will give up to two gallons (9 litres) a minute. These will need 22mm (or ¾in) diameter feed pipes.

Bath shower Before installing a shower over the bath, you will have to give some thought to the wall coverings around it, since these need to be waterproof to a height of at least 300mm (1ft) above the shower head. Either use ceramic wall tiles or cover with a waterproof plastic finish such as melamine-faced hardboard. You should also fill the join between the bath and the wall with a flexible setting silicone rubber sealant.

Finally, you are advised to fit a screen or plastic shower curtain round the open side of the bath and this should hang inside. Proprietary fittings are available here.

Once installed the shower will almost certainly be used more than the bath and this will create a lot of condensation, particularly if the bathroom is fully tiled. You can alleviate this problem by finishing the rest of the walls with, for example, pine panelling or cork tiles – some material that is warmer to the touch to reduce the possibility of moisture forming.

The best way to cut down on condensation, however, is to fit an extractor fan, ideally over the shower.

The simplest type of extractor fan to install is the window-mounted design, which fits into a hole cut in a pane of glass. Most glass suppliers will cut this hole to size for you. Usually this type of fan will incorporate a cord-pull switch.

The problem with this type, however, is that it would not normally be in the best position to extract the steam and moist air from a shower. The best solution here would be to fit a wall or ceiling-mounted fan. If this was combined with a tunnel or duct from the shower booth to the fan, the arrangement would create ideal extraction conditions.

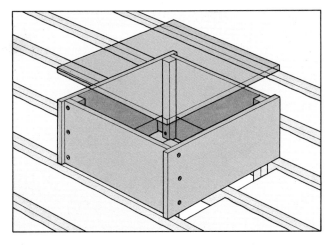

When raising the cold water storage cistern, you must use blockboard and not chipboard panels to ensure the platform is strong enough to hold the considerable weight of the cistern.

Remember that the fan must be wired into a fused connection unit and can only be operated by a cord-pull switch. Manufacturers of extractor fans supply detailed fitting instructions.

You must check before you fit the shower that there will be a sufficient head of water. This means ensuring that there is a minimum vertical distance of 1m (3ft) between the shower head and the bottom of the cold water storage cistern. If there is not sufficient head, there are two ways of overcoming the problem.

The cheapest method is to raise the cold water storage cistern in the loft. This will involve draining the cistern, disconnecting all the plumbing to it and providing a platform (see diagram) to give the necessary increase in height. Bear in mind that the platform must be strong enough to hold the weight of a full cistern.

Having drained the cistern and the rising main (see page 36), you will have to cut through all the pipes feeding to and from the cistern – that is the rising main, the hot and cold water feed pipes and the overflow pipe – and extend them by the required amount.

When cutting through the water service pipes, you should use a wheel-type pipe cutter to prevent the possibility of swarf from the cut edges getting into the pipes and damaging taps or valve washers.

You will need two straight pipe connectors (either capillary or compression type) and the required length of extension pipe for each of the service pipes.

If you have to raise the cistern, it is a good opportunity to check whether there are any signs of damage or wear in the cistern. If there are, then you should take out the old cistern and plumb in a new plastic one. And it is a wise precaution to replace the connection fittings and the ball valve at the same time.

The other method of increasing shower pressure is to fit a booster pump, which is electrically operated. You can either buy a dual pump, which boosts both the hot and cold water supplies before they are mixed, or a single pump, which boosts the mixed supply. The dual pump provides more efficient mixing as well as increasing the water pressure.

Do not forget that the switch for this pump must be cord-operated, since no standard switch is allowed in the bathroom. And the pump switch must be located outside the shower area. Incidentally, the pump must be wired to a fused connection unit and not to a standard 13amp socket outlet.

The mixer valve can be either manual or thermostatic. With the manual type, you have to set the controls to give the required rate of flow for both hot and cold water. This valve only controls rate of flow.

Bear in mind here that the temperature of the shower water can fluctuate as the level in the cold water storage cistern drops and the temperature of the hot water changes as it is used up. If you do fit this manual type, you must make sure that the maximum temperature of the hot water is at a level below scalding point in case the hot control only is turned on by mistake.

A good thermostatic mixer control has a safety device built in to prevent this happening. The valve controls the temperature of the outflowing water and adjusts accordingly should there be any fluctuation in the temperature and flow of the hot and cold water supplies.

If you live in a hard water area, check that the valve is designed to cope with this. This type of valve is designed not to be affected by scale formation and to remove scale as it is operated. Other valves are likely to be affected by scale to the extent that they lose their thermostatic function and will only operate as a manual control.

You can either plumb in the supply from the mixer valve to a fixed shower head as a permanent arrangement or use a flexible hose to an adjustable shower head. The latter type enables you to alter the height of the shower, depending on whether it is being used by an adult or a child, and means you can use it to wash your hair without taking a full shower.

As far as the supply to the shower unit is concerned, the hot and cold water should both be fed through at low pressure. This means taking a separate cold feed from the cold water storage cistern and a separate hot feed from the vent pipe above the hot water storage cylinder.

This typical shower fitting over a corner tub in the bathroom is operated by a thermostatic valve, which controls both the temperature and flow of the water through the shower.

This is to eliminate the risk of sudden changes in temperature caused by someone using another plumbing fitting elsewhere in the house (see pages 92–94).

Most shower connections accept 15mm (or ½in) diameter pipe. If, however, the water pressure at the shower head is borderline, it is a good idea to use 22mm (or ¾in) diameter pipe to increase the flow. To convert from the 22m (or ¾in) pipe to the 15mm (or ½in) shower outlet, you can get reducing fittings.

When arranging the pipe runs, make sure you bend the pipe round corners. And do not use elbow fittings, to avoid any possible restriction to the flow of water.

If you are installing a flush-mounted shower unit that fits into the wall, you should leave some means of access to the fitting in case of maintenance or repair work later. The ideal solution is to include a detachable panel in a suitable position on the wall. This can be bought.

When you have completed the installation of your shower unit, following carefully the manufacturer's instructions, adjust the mixer valve to the desired temperature to prevent the possibility of scalding. Stop washers are usually provided for this purpose. These washers vary with models, but usually they are fitted onto a serrated spindle with tags to prevent rotation past a certain point.

Instant electric shower This type of shower is designed as a self-contained unit that requires just a cold water supply. It must be fitted to comply with local water and electricity authority requirements, so check on these before you buy.

The cold water supply is normally fed direct from the rising main since most of these showers require a minimum water pressure of 10lb/sq in

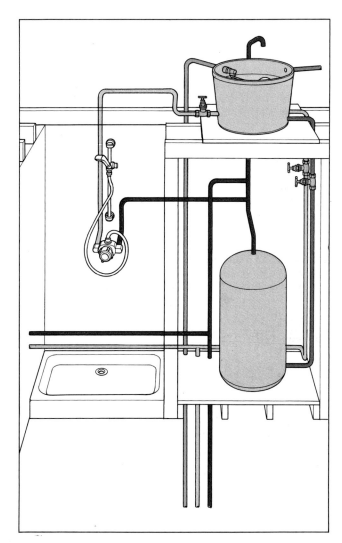

When plumbing in a shower, a separate feed is required for both the hot and cold water to ensure they are supplied at the same low pressure. The minimum head is 1m (3ft).

(equivalent to a head of water of 7m or 23ft) in order to operate satisfactorily.

The electricity requirement is heavy – between 4 and 8kW – which means you will need a separate fused supply direct from the consumer unit and must fit a double pole cord-operated ceiling switch.

For these reasons, this type of shower is not really suitable for DIY installation and you are therefore advised to have it fitted professionally.

With this type of shower unit the water temperature is dependent on the flow of water. This means that the water gets hotter as the flow rate drops – and the rate of output is therefore usually lower than with conventional showers. With a 7kW unit the output at a comfortable shower temperature is about two-thirds of a gallon (3 litres) per minute.

If you decide to buy an instant shower, make sure it is one that is recommended by your local electricity board and water authority and conforms to BS 3456.

Shower booth This separate unit is ideal where you want extra facilities in the bathroom or need to economise on space and replace the bath – or if you want to provide facilities elsewhere in the house. There are some major considerations, however, before you can go ahead with installation.

You must check that suitable arrangements can be made for the waste pipe (see chapter 6). This pipe should normally be of 40mm (or 1½in) diameter and run from a suitable trap below the shower tray at the minimum allowed gradient for no more than the maximum distance to either the main stack pipe or a convenient outside gully.

You will also have to find a suitable location for the unit if it is going to be fitted outside the bathroom. A large landing, disused fitted bedroom cupboard or an alcove may appear convenient, but the atmosphere around the unit will be damp and steamy during and after use and this may affect the surrounding areas. Equally a degree of privacy will be required by the person using the shower. And, of course, there must be reasonable access to a suitable supply of hot and cold water.

A large number of proprietary shower cubicles are available – either fully assembled or in kit form. These may contain their own shower control valves and head. A cheaper alternative is to buy a suitable shower tray and build your own enclosure.

You can buy a three-sided tray, two sides of which are at 90 degrees to each other and the third curved. This design fits neatly into a corner and you can enclose the front of it with a

The advantage of an instant electric shower is that the unit only needs a cold water supply. However the temperature of the water will depend on the rate of flow. This is not recommended as a DIY installation.

shower curtain. You can line the walls with any melamine-coated surface and seal the join of the tray with the wall using a flexible silicone rubber sealant. If you decide to put up a glass-sided booth or glass screen, make sure you use safety glass to BS 6262.

When locating the shower head, you should put it to one side of the booth so it does not aim water at the screen or curtain and therefore avoids unwanted drips and splashes.

Remember that the floor area around the shower should be covered with a water-resistant material and check that there are no electrical fittings or connections anywhere nearby or underneath the shower. From a safety point of view, you should also make sure that all electrical fittings or switches in the bathroom – or where the shower is fitted – are out of reach from the shower.

A small point, but worth bearing in mind, is that you will need a hook or peg close to the shower on which to hang a dressing gown, towel, etc.

The choice of shower itself is up to you. Whichever one you buy, make sure you follow the manufacturer's instructions carefully, bearing in mind the points already mentioned.

Installing a bidet

Although this fitting is commonly used in Europe, it has only recently gained popularity here. When fitted in the bathroom with a shower unit, it offers suitable facilities to enable you to dispense with a bath.

Space may be the major problem when it comes to installation and you should check that you are maintaining the recommended requirements around the various fittings in the bathroom (see pages 94–96).

Special three-sided trays are available to enable you to install a shower neatly into a corner. Screen off the unit by fitting a curved curtain rail above the front edge. Make sure the surrounding floor is waterproof.

If your house has a single-stack drainage system, the waste from the bidet should discharge into the main stack pipe. With the two-stack system, you must treat the waste pipe from the bidet in the same way as for a bath or wash-basin – that is, fed to the waste water pipe or over an outside gully.

There are two basic types of bidet and each involves different plumbing arrangements.

The most common type incorporates a simple over-the-rim water supply, which in effect makes the bidet a special purpose, low-mounted wash-basin. The water is fed into a mixer valve, which fits into the single hole at the back. The mixer may have an adjustable nozzle which provides a gentle stream of temperature controlled water for washing or can simply be used to fill the bowl.

With this type you can take the hot and cold water feed from the supply pipes to the wash-basin and bath. All bidets require 15mm (or $\frac{1}{2}$in) diameter supply pipes and the hot and cold water must be supplied at a balanced low pressure.

Because of the nature of its use, the bidet is fitted with a 'pop-up' waste system, whereby the drain plug is controlled by a knob in the middle of the mixer valve.

The other, more expensive style of bidet is known as the rim type with ascending spray. This has three holes at the back. The outer two are for the hot and cold control valves and the middle one for the spray and 'pop-up' waste controls.

With this bidet the water enters under the rim in a similar way to the water in a washdown WC pan and can either fill the bowl or warm the rim to a comfortable temperature. When you operate the spray control, a fountain of water provides the necessary rinsing facility.

To avoid the risk of contamination to water in the rest of the system from this type of bidet, certain conditions have to be complied with. The base of the cold water storage cistern must be at least 2·7m (or 9ft) above the rim of the bidet. The cold water supply must be through a separate 15mm (or $\frac{1}{2}$in) diameter pipe running direct from the cold water storage cistern and the hot water supply must also be through a separate 15mm (or $\frac{1}{2}$in) diameter pipe running direct from the vent pipe above the hot water cylinder.

As far as the plumbing is concerned, the fitting of a bidet is very similar to that of a wash-basin (see pages 103–104). The bidet itself must be screwed firmly to the floor to prevent any movement which could otherwise damage the joints in the supply and waste pipes.

The arrangements for fitting the waste, including the 'pop-up' control, do differ according to the model, but manufacturers supply detailed instructions and you should follow these carefully. You should certainly use mastic in the join between the waste and the bowl.

You can fit bottle traps to the bidet waste pipe, which should normally be of 35mm (or $1\frac{1}{4}$in) diameter. If space permits, the deep 75mm (or 3in) seal type trap is recommended.

Installing a bedroom vanity unit

In the average modern house where there is only one bathroom, which often includes the only WC as well, problems are often experienced when demand for the facilities increases – particularly early in the morning when the family are getting up.

One way of alleviating the congestion is to fit a wash-basin in one or more of the bedrooms as

This over-the-rim bidet has a single hole connection for the mixer valve, with a pop-up waste control combined with it.

This ascending spray bidet has three holes at the back. The outer two take the hot and cold water taps, while the centre one is for the combined pop-up waste and spray control.

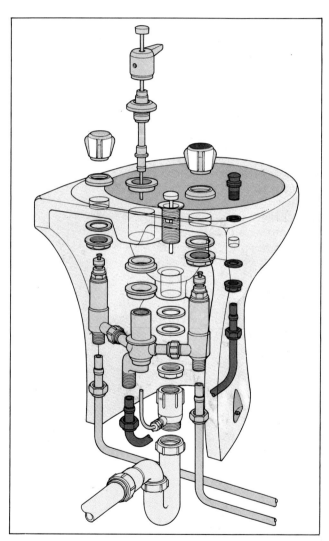

This shows the connections for a three-hole bidet with ascending spray. You will need separate hot and cold feeds at low pressure with a minimum head of 2.7m (9ft).

well – a job that is neither too difficult or expensive, provided that when you choose the siting for this fitting you take into account certain plumbing requirements.

The waste pipe run from the wash-basin must be as straight and as short as possible. Obviously you will want it to be unobtrusive, but you should try to restrict the run to 1·7m (6ft 8in) from the wash-basin to the main stack pipe. It is possible, however, to accommodate a longer run if there is no alternative (see pages 95-96). Or you can organise to have the waste discharging into an outside gully, if this is more convenient.

From a safety point of view, you must check that the position of the wash-basin does not clash with electrical installations around or beneath it and that there are no switches in reach. The other major considerations include the water supply and here you will want to keep the wash-basin as near to existing pipe runs as you can to reduce the amount of plumbing needed.

You will also need to waterproof the area around the wash-basin – that is, the wall and the immediate floor space.

From the conditions already described it is obvious that the most convenient place to site an extra wash-basin will be on a wall adjoining the bathroom. Another potentially suitable location would be in a disused fitted cupboard, since you can then keep the wash-basin out of sight when not in use.

There are several proprietary vanity units available, although it is not too difficult to make one to suit your needs. This could involve a simple timber frame construction with a top made of melamine-covered chipboard. You can get this board in white or a wood grain finish, although you can fit a suitable melamine design with impact adhesive onto 15mm ($\frac{5}{8}$in) thick chipboard to tie in with the room decor. You then need to cut out a suitable size hole in the surface with a padsaw to take the wash-basin. The normal fixing height for a wash-basin is 800mm (or 32in) from the rim down to floor level.

Make sure when you choose a wash-basin that it is not too small. Since it will be used for 'top half' washing, it needs to hold a reasonable amount of water.

The inset style of basin is available in ceramic, plastic, glass fibre or stainless steel and normally the fitting will be supplied with all the fixings necessary.

If you are making up your own unit, ensure that the join between the wash-basin and the surrounding surface is watertight. You should use non-setting mastic to achieve this.

The type of tap you fit is a matter of choice, although if you want a single stem mixer tap, make sure you get a wash-basin with just one hole in the back. If you buy the shrouded-head style of tap, you can fit different plastic handles later on if you change the colour scheme in the room.

You should make sure the waste is discharged through a deep seal bottle trap and, if you have to fit a long run of waste pipe, you would be advised to install an anti-siphon trap. The normal waste pipe for a wash-basin is 35mm (or $1\frac{1}{4}$in) diameter.

You must waterproof the wall behind the wash-basin or vanity unit. Here you have plenty of choice, too. You can fit melamine-faced hardboard to that section of the wall or fix ceramic or mirror tiles or a single tile mirror. Whatever type of covering you use, make sure you seal the join with the wash-basin or unit with a silicone rubber sealant.

Index

Figures in italics refer to illustrations.

baths 11, 96
   replacing 100, *102*, 103
bathroom fittings, replacing 94–6
   space required for *94*, 95
bending pipes 22, 26–8, *47*, 48
bidet, installing 114–116
blowtorch 22, 23, *42*, 49
boiler 12, 14, 15, 16, 25, 43, 81

calorifier, sealed coil 16, 17, 25, 36
capillary fittings 22, 45, 48–50, *49*, *50*
   end feed 48, 49
   fitting 48–9
   leaking 50
   repairing 37
   Yorkshire 48, 49
central heating system 12, 15, 16, 43
ceramic discs 24, *24*, 55
cesspool 17
chassis punch 23
cistern *see* storage cistern
compression fittings 20, 45, 48, 50–2, *51*, 60, 63, 70, 77, 83, 85
   fitting 50–2, *51*, 63, *63*
   on plastic 76, *76*
   repairing 37–8
corrosion 9, 10, 12, 13, 16
   inhibitors 15, 16
cylinder *see* storage cylinder

dezincification 9–10
dishwasher 77
   plumbing in 80, 88–90
drainage systems *see* waste systems
draincock 11, 15, 36, 80, 81, 107
draining the system 15, 25, 36, 37, 80, 81, 98, 100

earth return, electrical 8, 58, 84
electricity, turning off 43
emergency action for burst pipes 43
emery cloth, 31, 81
expansion tank 12, 13, 16
   capacity 12
extractor fan 110

feed tank *see* expansion tank
file 20, *23*
fittings, brass 10
   dezincification resistant 11
   plastic 65–6
   special 52–3
   see also capillary fittings, compression fittings
float arm, cistern 29, 31, 32, 33
   repairing 29
freezing 24, 29, 38, 39, 41, 43
frost damage 11, 12, 17, 36, 39
frost repair kits 42, *42*

gland nut 26, 27, 28, 65
glass fibre bandage 43
glasspaper 22, 42, 43, 62

hacksaw 20, *23*
hammer 20, *21*
hole saw 23

immersion heater 14, 15, 36, 81
   single element 15
   twin element 15
insulation 12
   see also lagging

jointing materials 20, *21*, 37, 52, 65, 81, 83, 92, 102
joints
   mechanical 63
   push-fit connectors 52
   remaking 42
   repairing leaking 36–8, *36*
   repairing plastic 38
   Acorn, push-fit 60, 63–5, *64*
   see also capillary fittings, compression fittings, ring seal joints, solvent-welded joints

lagging, cistern or cylinder 39–41, *40*
   materials 38, *38*, 39, *39*, 40, *40*, 41, *41*
   pipes *38*, 38–9, *39*, 43, 108
metal components 45–57
   reaction between 9

packing 27, 28, 36
   materials 20, *21*, 28
petroleum jelly 26, 28, 31, 32, 36, 44, 46
pipe bending machine 22, *23*, 45, 46

pipe bending spring 22, *23*, 46
pipe cutter 20, *23*
pipes 15
   annealing copper 46
   chromium-plated copper 46
   clearing blocked 43–4, *43*, *44*, 77
   clips 66, *66*, 67
   copper 9, 10, 11, 45, 65, 67, 69
   epoxy resin repair kit 42–3
   feed 11, 12, 13, 16
   frozen 43
   galvanised steel 8, 9, 10, 11, 46
   Kopex 52, 85
   large bore 11, 16
   lead 8, 11, 46
   nominal bore 10, 11
   old systems 8–9
   plastic 11
   repairing burst 41–3, *41*, 48
   replacing sections 42
   sizes 10, 45, 46
   small bore 16
   stainless steel 10, 45
   vent 13, 16, 17
   warning 12, 16
   see also tubing, waste pipes
plastic components 9, 58–66
pliers 19 *21*, 29, 30, 31
plunger *43*, 44
pressure failure 11, 12
primary circuit 15, 16, 36
pumps 14, 16

radiators 12, 15, 16
regulations, local authority 8, 11, 17, 105
   see also water authority
ring gland 28, 63, 65, 91
ring-seal joints 67, 68, *69*, 70, 71, 75, 76, 77, 103, 105
reservoirs 8, 11
rising main 11, 12, 107

safety 8, 60
screwdrivers 19, *20*
secondary circuit 16, 36
showers 56, 91, 105, *111*
   booth 96, 108–9, 113–4
   electric 112–3, *113*
   head required 93, *93*
   installing 92–4, 108–114

   over bath *109*, 108–9
   pumps 111
sinks, replacing 84–8, *85*, *86*, *87*
soldering materials 22, *23*
solvent-welded joints 60–2, *61*, 67, 68, 69, 71, 72, 74, 74–6, 88, 105
spindle 28, 55, 65, 91
   gland 27, 55
spanners 19, *21*
   adjustable 19
   compression joint 20, *23*
   crowsfoot 81, 92, 101
stopcock, company 9
   freeing jammed 35–6
   non-return 8
   main 8, 11, 25, 29, 35, 36, 81, 107
   water authority 11, 36
storage cistern, cold water 9, 11, 12, 13, 15, 16, 23, 25, 29, 31, 32, 43, 80, 89, 92, 109
   insulation 12
   lagging 39, *40*, 41
   raising *110*, 110–111
storage cylinder, hot water 12, 13, 15, 16, 17, 25, 56, 80
   lagging 41

tallow cotton 20
tank cutter 23
taps 11, 55, 55–7
   bib 24, 27, 36
   connector kit 52, 85, 102
   connectors 52, *53*
   conversion kit 52–4, *54*, 91
   dezincification resistant 11
   fitting 22, *56*
   fitting mixers 80–3, *82*, *83*, 92–4
   fitting outside 11, 107–8, *108*
   inspecting 25
   labelling 11
   mixers 55, 56
   plastic 65–6, *65*
   pillar 55
   reassembling 26, 28
   removing head 25, 28
   repairing dripping 24–7, *26–7*
   repairing leaking 27–8, *28*
   replacing bathroom taps 91–2, *91*
   shrouded head 24, 27, 55, 91
   sizes 11, 55

Supatap 24, 26–7, 55
types 24, *57*, 91
tools 19–23
traps 17, 43, 76–9, *78*, 85, 87, 104, 105, 106
   anti-siphon 77
   bottle 77–9
   brass 17
   clearing blocked 44, 79
   copper 17, 77
   lead 77
   polypropylene 77
   removing 44
   self-resealing 77–8
   tubular 79
tubing
   plastic 9, 58, *59*
   polybutylene 10, 58
   polypropylene 10, 67
   polythene 10, 60, 63
   PVC 10, 58
   see also pipes, waste pipes

valves
   ball 16
   conversion sets 66
   Croydon 29
   dezincification resistant 11
   diaphragms, replacing 32, *32*
   float-controlled 12, 29, 31
   gate 12, 16, 25, 29, 35, 43, 55, 81, 92
   Garston 29, 33
   Portsmouth 29
   repairing cistern 29–33
   repairing Garston 31–2, *31*, *32*
   repairing Portsmouth 29–31, *30*
   repairing Torbeck 33, *33*
   safety 14
   Torbeck 33, *33*
vanity unit, installing 115–117, *116*

wash-basin, replacing 103–4, *104*
washers 24, *24*, 25, 30, 55, 63, 81, 85, 91, 93, 101, 104
   replacing 12, 24–7, *26–7*
   replacing Supatap 26–7
washing machine, plumbing in 54, 80, 88–90, *89*, *80*
   plumbing in kit 88, *88*
waste and overflow kit 85, *85*

waste pipes 17
   angle of 68, 71, 84, 88, 95, 102, 105
   bath 68
   bend fittings 69, 70
   clips for 69, 71
   copper 67
   expansion of 68–9
   fittings for 67, 68, *68*
   joining 74–6
   length of 88, 95, 105, *106*
   lead 67
   muPVC 67, 68, *68*, 69, 72
   overflow 29, 67, *67*
   plastic 67
   polypropylene 67, 68, 69, 70, *70*
   sink 68
   sizes 67
   soil pipe 17, 70–1, *71*, 72, 95, 97, 98, 100, *107*
   stack pipe 17, *73*, 84, 87, 88, 90, 93, 105–7
   uPVC 67, 71
   wash-basin 68
waste systems 11, 17–18
   single-stack 17, *18*, 77
   two-stack 17, *18*, 77
water authority 8, 9, 11, 58, 71, 80, 105, 113
water, daily requirements 7
   hard 8, 9, 15
   heating systems 25, 36
   soft 8, 9
water supply 8, 9, 11, 12, 16
   cold distribution system 11–13, *13*
   direct 14–15, *14*
   hot distribution system 13, 14–17
   indirect 14–15, *15*, *16*, 16–17
   turning off 12, 25
WC, cistern 11, 29, 31, 34, *34*, 48
   close-coupled 95, *95*
   connections 17, 72, *72*
   replacing 96–100, *99*
   replacing cistern diaphragm 33–5, *34*, *35*
   single to dual flushing 34–5
   syphonic 96–7, *97*
   washdown 96, *97*
   waste fittings 71
wrench, basin and bath 22, *23*
   self-grip 20, *21*, 25, 32
   Stillson 19, 20, *23*